THE QUESTION OF ANIMAL AWARENESS

Evolutionary Continuity of Mental Experience

THE QUESTION
OF ANIMAL AWARENESS

Evolutionary Continuity of

Mental Experience

DONALD R. GRIFFIN

Revised and Enlarged Edition

THE ROCKEFELLER UNIVERSITY PRESS

NEW YORK · 1981

COPYRIGHT © 1981 BY THE ROCKEFELLER UNIVERSITY PRESS

LIBRARY OF CONGRESS CATALOGUE CARD NO 81-51221

ISBN 87470-035-3

PRINTED IN THE UNITED STATES OF AMERICA

REVISED EDITION 1981

Contents

239652

Preface

The "ferment of constructive excitement" in ethology mentioned in the preface to the first edition of this book has continued unabated, rekindling serious scientific consideration of the degree to which nonhuman animals think consciously about objects and events, and about themselves. It had often been tacitly assumed that no other species shares with us the capacity for conscious thinking, planning of future actions, or any of those mental experiences which are known under the general term "awareness." The correctness of this assumption, or the degree to which it requires modification, is crucial to our understanding of the human condition. Just what is and is not unique to our species? The difficulty of this question had led to its neglect by behavioral scientists for several decades, with the result that we have little firm evidence to guide our judgments or constrain our speculations. But ethologists and psychologists have now begun to inquire how, if at all, it may be possible to gather objective and verifiable data about animal awareness.

This renewed interest is well represented in two important publications which appeared in 1978. The first is a symposium volume entitled *Cognitive Processes in Animal Behavior*, edited by Hulse, Fowler, and Honig. Athough several chapters present data suggesting awareness, the authors do not directly confront the issue of animal consciousness. The second is an entire issue of *The Behavioral and Brain Sciences* (Vol. 1, No. 4) devoted to "Cognition and Consciousness in Nonhuman Species" (Premack and Woodruff, 1978; Griffin, 1978b; Savage-Rumbaugh et al., 1978). This new scientific journal contains general articles together with multiple commentaries by interested scientists and scholars and responses from the authors. These extensive and thoughtful discussions have contributed greatly to clarifying my own ideas, and I have drawn on them extensively in preparing this revised edition, as well as on material I have published elsewhere since 1976 (Griffin 1977, 1978a).

One especially encouraging aspect of this renewed attention to the questions of animal thinking has been its multidisciplinary nature. Important ideas have been contributed not only by ethologists, but by evolutionary biologists, comparative psychologists, linguists, and philosophers. Since it would be foolish to overlook data and insights from any quarter, I make a special effort in this new edition to review the contributions of thoughtful scholars from a wide variety of fields.

Because the issues under discussion are so elusive and difficult, many of the ideas most worthy of our attention can best be appreciated in an author's own, carefully chosen words. Therefore I have included a number of direct quotations, some in the text and others in annotations of certain bibliographical references. (The annotated references are marked in the text by *.) I believe that I have avoided the danger of distorting any writer's meaning by using quotations out of context, and that any residual risk of such errors is more than offset by the great interest and significance of the questions at issue. In this way, readers can draw their own conclusions as directly as possible from the clearest statements of both the evidence and varying interpretations.

I am grateful to all those whose contributions were acknowledged in the Preface to the first edition, because they have continued their invaluable assistance and encouragement.

The second edition has also benefited from constructive criticisms by several colleagues from The Rockefeller University and neighboring institutions, especially Colin Beer, Susan Calhoun-Radano, Herbert Terrace, and Robert L. Thompson. But my principal source of critical support and active collaboration has been Carolyn Ristau, who has modulated many of my naive enthusiasms and helped me to clarify and strengthen important topics and arguments. She has been all the more helpful by *not* buying all of my ideas, and she should not be held responsible for anything with which a given reader may disagree.

In addition, I thank the following publishers for their permission to reproduce lengthy quotations: Cambridge University Press, *The Behavioral and Brain Sciences*, and Springer-Verlag, *Behavioral Ecology and Sociobiology*.

I wish to take this opportunity to express my gratitude to the Harry Frank Guggenheim Foundation for its generous support of research on cognitive ethology, and especially for supporting the preparation of this revised edition.

Preface

to the First Edition

A ferment of constructive excitement is evident in ethology, the study of animal behavior with emphasis on evolutionary adaptations to the natural world. For example, social organization, individual recognition, altruistic behavior, endogenous activity rhythms or biological clocks, and complex systems of orientation and navigation have been identified in more and more species not previously suspected of having such complications in their ways of life. All scientific discoveries contain some element of novelty, but ethologists now feel confident in making statements that differ qualitatively from anything that was scientifically thinkable forty or fifty years ago. Since there is no reason to believe that this progress will suddenly come to a halt, it is worthwhile to outline some directions in which ethology may develop. In this attempt at speculative extrapolation, it is especially appropriate to pose some new questions and to reopen certain old ones from a fresh perspective. Most of these questions relate to the general issue of our evolutionary kinship to other species of animals, with special reference to the more complex cognitive functions that appear to regulate the behavior of animals and men.

Many thoughtful colleagues, individually and collectively, have provided essential stimulation without which this book would never have been written. Most important has been the stimulating environment òf The Rockefeller University, which provided the opportunity for serious and consistent concentration on distant objectives. Thomas Nagel of Princeton University supplied an immediate spur while visiting our campus when he raised the question of whether animals have mental experiences. Peter Marler and Fernando Nottebohm, along with many other colleagues, offered invaluable and always constructive criticism. The "negative feedback" from my colleagues has been just as reinforcing as the positive encouragement. Rosanne Blair has patiently toiled over the preparation of innumerable draft manuscripts. Helene Jordan of The Rockefeller University Press has been a most perceptive editor. Finally, Jocelyn Crane supplied inestimable general encouragement, together with abundant factual knowledge from her wealth of experience with the real, natural world of animals and their ethology.

It is a pleasure to extend my grateful acknowledgments to the following publishers for permission to reproduce substantial quotations:

American Philosophical Association, *Thoughtless Brutes;* Braziller, *Signs, Language, and Behavior;* Clarendon Press, *Essays on Function and Evolution in Behaviour;* The Free Press, *Words and Things;* Harcourt Brace Jovanovich, *Language and Mind;* Harper & Row, *Cartesian Linguistics* and *Language: An Inquiry into its Meaning and Function;* Johns Hopkins Press, *Mind, an Essay on Human Feelings,* Vols. I and II; Macmillan, *Philosophical Investigations;* Oxford University Press, *Bowerbirds, Their Displays and Breeding Cycles;* Popular Library, *Supermoney;* Random House, *Psychological Explanation, An Introduction to the Philosophy of Psychology;* Teachers College Press, *Sciences, Psychology and Communication;* Wiley, *Learning Theory and Behavior* and *Learning Theory and Symbolic Processes;* Yale University Press, *The Philosophy of Symbolic Forms.*

I am also grateful to the following colleagues for permission to read and to quote from books or articles in preparation or in press: Colin G. Beer, Roger S. Fouts, James L. Gould, Peter Marler, Duane M. Rumbaugh, and W. John Smith.

1

Expanding Horizons
in Ethology

Ethologists and comparative psychologists have discovered a host of refined adaptations in animal behavior during the past few decades. Food-finding, avoidance of predators, and behavioral adaptations to environmental stresses, including constructing shelters, nests, and burrows, all involve impressively versatile tactics on the animal's part, rather than rigid, stereotyped reflexes. Social behavior, especially courtship and care of developing young, call forth an efficiently tuned and controlled matrix of interactions among many different and potentially conflicting behavior patterns. Animal orientation and navigation have provided several striking examples of previously unsuspected modes of perception. Finally, the versatility of animal communication used to coordinate group activities has implications that can only be described as revolutionary.

The flexibility and appropriateness of animal behavior suggest both that complex processes occur within their brains, and that these events may have much in common with our own conscious mental experiences. To the extent that this proves to be true, many of our ideas and opinions about the relationship between animals and men will require modification. The current scientific *Zeitgeist* almost totally avoids consideration of mental *experience* in other species, while restricting attention to overt and observable behavior and physiological mechanisms. To the extent that animal thinking and feeling become accessible to scientific scrutiny and analysis, ethology will be greatly broadened and enriched. The implications of these developments are profound. For one thing, they oblige us to reconsider deep-seated assumptions about human nature, and to inquire whether our kinship with other living organisms may be closer than we have been accustomed to recognize. Some of those mental attributes which we have been accustomed to view as unique prerogatives of our species may turn out to be more widely distributed, although presumably in limited and simpler forms. If so, it becomes reasonable and promising to attempt the study of mental experience in animal surrogates. This book will therefore examine both the pertinent evidence and its general significance in the hope of stimulating renewed interest in, and investigation of, the possibility that mental experiences occur in animals and have important effects on their behavior.

Because the available data are far from adequate, tentative speculations must serve as first steps toward future investigations. It will be necessary to raise or reopen difficult, fundamental questions, many of which have been neglected for some time. I will also try to relate such questions to previous discussions and to realistic possibilities for experimental investigations in the future.

Communication with One's Environment

Scientists often have found that very large and important questions can best be tackled in a gradual and piecemeal fashion, by nibbling at their edges, so to speak, before plunging into their central core. As explained in later chapters, I believe that the communicative behavior of animals offers especially promising insights into whatever thinking may be going on in their brains. But the practical difficulties impeding clear-cut experimental analysis are so formidable that it is wiser tactically to begin with more manageable material, which nevertheless has produced the same sort of surprising discoveries and extensions of our patterns of expectation as have the more complex and subtle examples that are discussed later. My own interest in animal consciousness has been aroused and sustained throughout many investigations of the orientation and navigation of bats and birds, because it has seemed intuitively that the animals I was studying may have known what they were doing in much the same way we might realize what we were about if engaged in comparable behavior.

The orientation of an animal can be viewed as a process of communication with its surroundings, in the sense that very weak signals from the environment trigger behavioral responses which release much greater physical energy and produce important biological results. Usually, but not always, pertinent information flows in only one direction—from the environment to the animal's sense organs. As in true communication between individual animals, selective attention is paid to particularly useful physical signals, usually only one or a very few out of many available to the animal. It has sometimes been difficult to identify the environmental signal that is actually selected, and a few specific examples illustrate how discoveries about animal orientation have opened our eyes.

Orientation behavior provides several instructive cases in which scientists have tended to minimize the versatility of animals, and in which later investigations demonstrated much more refined and flexible behavior than had seemed at all plausible a few years earlier. For example, as a naive student in the late 1930s, I and others toyed with the notion that birds might orient themselves by the sun or stars during their homing and migratory flights. But my elders and betters were emphatic in discouraging what seemed to them—and, on more sober reflection, even to me—a rather silly and romantic line of speculation. "Why, the poor birds would need to carry around a whole set of tables, a sort of almanac, to correct for the motions of the sun and stars across the sky." In those days, no respectable biologist felt comfortable with anything more complex than orientation toward, away from, or perhaps at a fixed angle to a source of light. Animal behavior was treated as a set of tropisms and taxes, as reviewed by Fraenkel and Gunn (1961). A sort of simplicity filter shielded us from worrying about possible complexities. It was even quite possible to explain most of the data on the homing of birds on the basis of random wandering (Wilkinson, 1952) or exploration (Griffin, 1952).

In the 1950s, however, Matthews (reviewed in Matthews, 1968), Kramer (1959), and Sauer (1957) showed that birds are fully capable of making at least approximate corrections for the motion of the sun, and even of the stars, across the sky. Birds do indeed practice time-compensated sun- and star-orientation (Schmidt-Koenig, 1965; Emlen, 1967, 1975; and several papers in a symposium edited by Schmidt-Koenig and Keeton, 1978). Furthermore, improved homing experiments showed that when pigeons are carried into unfamiliar territory in directions arbitrarily selected by the experimenter, they often show goal-directed homeward orientation within a minute or two after release.

Similar simplicity filters held back my thinking for several years in another area. Robert Galambos and I discovered echolocation in bats with the aid of then-unique electronic apparatus developed by the physicist G. W. Pierce (reviewed by Griffin, 1958). Pierce's apparatus was capable of detecting sounds above the frequency range of human hearing, and when I first brought bats to his laboratory it was obvious that these animals emitted ultrasonic sounds almost continuously. But when bats were allowed to fly near the apparatus, these sounds were only occasionally detectable. I did not realize at the time how directional Pierce's equipment was; that is, how greatly its sensitivity was reduced when bats were not close to the axis of the parabolic horn in which the microphone was mounted. I therefore suspected that these newly discovered sounds might simply be call notes, not necessarily used for orientation. This conservative error was soon corrected after more detailed observations and experiments, but it is a significant example of the dangers inherent in limited imagination when one is dealing with new and unknown phenomena. Our experiments soon confirmed one of several speculative explanations, that of Hartridge (1920), for the ability of bats to fly without collisions through the complete

darkness of caves by emitting sounds above the human frequency range and hearing echoes from obstacles.

Despite the excitement of solving this long-standing mystery, and despite the opportunity to measure the acoustical
properties of the orientation sounds by which bats avoid obstacles (Griffin, 1946, 1950), it was several years before my thinking
progressed to the point of seriously wondering whether bats
might also use echolocation in hunting insects. Authoritative
opinion warned that tiny flying insects would not return enough
acoustical energy to yield audible echoes, and the whole idea
simply seemed too farfetched for serious consideration. Yet
here, again, the zoological reality turned out to exceed my initial
speculations (Griffin, 1953, 1958, 1980). When I first took the
trouble to observe wild bats hunting insects, it became clear
that during insect pursuits they increase the repetition rate of
their ultrasonic orientation sounds more sharply than had ever
been observed in the laboratory. Such increases in repetition
rate accompany the detection of small obstacles and also occur
when a bat prepares to land. This was strong suggestive evidence that insects were detected by echolocation.

Conclusive experimental evidence was not obtainable for
several years until we learned how to elicit insect-hunting behavior in relatively small rooms where controlled experiments
were feasible (Griffin et al., 1960). Little brown bats *(Myotis
lucifugus)* managed to capture fruit flies at rates of several per
minute, and we could measure their hunting success by weighing the bats before and after periods of intense feeding activity.
Reluctance to believe that bats could successfully pursue and
capture tiny insects by echolocation had led to the alternate
explanation that their pursuit maneuvers were guided not by
echoes from the insects, but by localizing and homing on the
sounds of insect wingbeats. But in a darkened room filled with
loud audio-frequency noise, which completely masked the faint
flight sounds from the fruit flies, bats gained weight at essentially the same rate as they did in the quiet with lights on. In
other experiments with relatively weak ultrasonic noise, the

bats ceased all attempts to capture flying insects. More recent experiments have demonstrated that bats have a highly refined capability for detecting faint echoes and also for discriminating among different classes of echoes according to their timing and frequency spectrum (Griffin, 1973; Simmons et al., 1975). Echolocation by insectivorous bats is of sufficient importance under natural conditions that some groups of insects have even evolved auditory receptors sensitive enough to ultrasonic frequencies to warn them of approaching bats (Roeder and Treat, 1957; reviewed by Roeder, 1970; Fenton and Fullard, 1979).

Not only bats, but whales, porpoises, and dolphins use echolocation both for general orientation and to capture moving prey (reviewed by Kellogg, 1961; Norris, 1966; Griffin, 1958, 1973; Popper, 1980). Two genera of cave-dwelling birds find their way to nests that are sometimes deep inside caves where it is totally dark (the oilbirds of South America, *Steatornis caripensis*, and swiftlets of the genus *Collocalia* or *Aerodramus* in southeast Asia and Australia). Their echolocation is based on clicks that are clearly audible to human ears, and some of the swiftlets can detect obstacles as small as quarter-inch rods (Griffin and Suthers, 1970). The whole subject of echolocation as it has evolved in both bats and marine mammals has been thoroughly analyzed in several chapters of a symposium volume edited by Busnel and Fish (1980).

A comparable surprise was the discovery that electric fishes orient themselves by sensing changes in the fields produced by their own electric organs (Lissmann, 1958; Lissmann and Machin, 1958). This initial discovery of a wholly unsuspected new sensory modality has been followed by detailed investigations of the neurophysiology of electroreception, of the variety of signals used by different kinds of electric fish, and of the use of similar signals for social communication (reviewed by Bullock, 1973; Heiligenberg, 1977; Hopkins, 1974, 1977, 1980).

Still another unexpected discovery was the ability of honeybees to compensate for the motion of the sun through the sky and to orient themselves by the polarization patterns of the blue sky (Frisch, 1950, 1967). Sensitivity to polarized light has since been demonstrated in many other arthropods and fish (Waterman, 1974), and in some individual homing pigeons by Kreithen and Keeton (1974a).

Important as all these discoveries were, they do not, of course, mean that all speculations about animal behavior will eventually prove to be correct. It may well be that, during the past generation, research on orientation behavior has disclosed a disproportionate share of such unexpected developments. Nevertheless, it is reasonable to inquire whether there is any end to this series of surprises. A contemporary case in point is the evidence that birds can sense and orient by the magnetic field of the earth (discussed extensively in the symposium edited by Schmidt-Koenig and Keeton, 1978). The positive evidence is limited so far to inconsistent and short-duration effects of magnets on initial headings of homing pigeons (Keeton, 1974; Walcott and Green, 1974) and slight shifts in the directional orientation of migratory birds in small cages (Wiltschko, 1974, 1975)— shifts so small that they are suggested only by statistical analysis of hundreds of responses. The situation is complicated by the fact that several attempts to demonstrate consistent and unequivocal responses of birds to weak magnetic fields have been totally unsuccessful (Kreithen and Keeton, 1974b; reviewed by Griffin, 1978c). On the other hand, stronger evidence for sensitivity to the earth's magnetic field has been reported in fish by Kalmijn (1971, 1974, 1978) and in honeybees by Lindauer and Martin (1972) and by Martin and Lindauer (1973).

The most convincing of all evidence that living organisms can orient to the earth's magnetic field is the discovery by Blakemore (1975) that certain mud-dwelling bacteria contain iron in the form of magnetite, and that this enables them to swim toward magnetic north. This directional orientation can be altered experimentally by an artificial magnetic field. J. L. Gould and his colleagues have discovered small amounts of magnetite in honeybees and homing pigeons (Gould et al., 1978; Walcott et al., 1979; reviewed by Gould, 1980). This has not yet been shown to be used in orientation, but its presence is at the very least highly suggestive. We may soon, though we do not yet, have truly convincing evidence that animals have a sensory window on the earth's magnetic field.

Close study of previously unexplained capabilities for orientation has thus revealed unsuspected sensory channels. In each case, a particularly appropriate physical signal from the environment is utilized selectively by a specialized sensory system. Echolocation and the electrical orientation of certain fishes are exceptional, because they are active processes during which the animal literally questions its surroundings with probing signals that have been adapted for this function through a long evolutionary history. Bats and porpoises both make rapid and appropriate adjustments of the motor mechanisms of sound emission and of neural networks in the brain, thereby adapting the entire process of echolocation to the avoidance of stationary obstacles, searching for insect prey, or the pursuit and interception of flying insects. These reactions include increases in the repetition rate of orientation sounds when the animal faces a difficult problem, and in bats there are also changes in the duration and pattern of frequency sweep used in each orientation sound. Changes in the sensitivity of peripheral portions of the auditory system within a bat's brain occur a few thousandths of a second after each orientation sound is emitted. These neural adjustments are superbly adapted to enable bats to detect faint echoes returning from objects at varying distances (reviewed by Suga and O'Neill, 1980).

Cognitive Maps

The capabilities for perceptual organization that an animal requires for complex orientation behavior include the establishment of what has been called a cognitive map by Tolman (1948) and Olton (1979) or "an elementary map of the environment" by Thorpe (1974b). When animals maintain appropriate orientation during long and challenging journeys by integrating a variety of environmental stimuli, their brains must include internal representations of the outside world. In discussing reductionism in biology, Thorpe points out that "problems of spatial position, of orientation and of direction finding . . . lead us naturally to the problem of conscious self-awareness."

The experiments of Menzel et al., discussed in Chapter 4, point to the presence of cognitive maps in chimpanzees. Tolman based his concept of a cognitive map on experiments with maze learning in laboratory rats, and more recently Olton and Samuelson (1976) demonstrated that rats can remember which of several arms of a radial maze they have already explored. The hippocampus of mammals contains neurons that are activated when the animal is in a particular place, regardless of its bodily orientation. This has suggested that the hippocampus serves in part as an actual cognitive map (O'Keefe and Nadel, 1978).

A suggestive example is provided by echolocating bats—not so much by their impressive successes, as by certain situations in which their orientation fails. When flying through thoroughly familiar surroundings, many bats seem to rely heavily on spatial memory. Although orientation sounds continue to be emitted in an apparently normal manner, the bats collide with newly placed obstacles, turn back from the former location of objects that have suddenly been removed, and even try to land on a familiar toehold that has been taken away (Moehres and von Oettingen-Spielberg, 1949; Griffin, 1958). Bats behaving in this way remind me of the collision of the *Andrea Doria* and the *Stockholm* in thick fog when, according to contemporary newspaper accounts, both ships were equipped with properly functioning radar sets. What I like to call "Andrea Doria bats" apparently pay attention only to their internal images of spatial relationships, even though echoes from the newly placed barriers reach their ears at far higher intensity levels than do those of small wires or insect prey which they detect readily under other circumstances. In any event, this kind of behavior clearly demonstrates that some sort of internal map or stored pattern representing the familiar environment must exist in the bat's brain.

The orderly complexity of cognitive maps in animals suggests, though it does not prove, that some of the time they may consciously think about their orientation. An animal might have a detailed internal map in its brain without being consciously aware of the geometry of its surroundings. But it is sensibly conservative to recognize that we simply do not know. These unexpected complexities have stimulated some ethologists to reconsider the general question of subjective mental experiences in animals. Renewed attention to these long-neglected, but highly significant, questions is now especially appropriate and timely, because recent advances in ethology have opened up new lines of experimentation which offer realistic hopes that these questions can be answered.

Inquiry versus Assertion

When scientists consider conscious mental experience, they exhibit an almost irresistible tendency to assert definite conclusions, even though the available evidence is crude and inadequate. This often involves a surprising inversion of the customary relationship between the definiteness of assertions and the quality of the available evidence, as schematically illustrated by Figure 1. Curve A represents the ordinary situation in scientific discussions where the firmness of convictions is at least roughly proportional to the validity of the available data. Curve B seems more descriptive of what one finds in discussions of the nature of consciousness or the degree to which other species share with us the capacity for any kind of mental experience.

FIGURE 1.   Diagram of the double standard often applied to ordinary scientific data (A) and discussions of animals consciousness (B).

Definitions

Terms such as mental experiences, mind, awareness, belief, intention, or consciousness are obviously difficult to define; and one reason for avoiding a cognitive approach to animal behavior has been confusion about the meaning of the terms and concepts involved. Depending on how awareness is defined, its possible existence in other species can vary all the way from being trivially obvious to the most preposterous level of implausibility. At the first extreme, one might define awareness as any capacity for reaction; but this would allow the inclusion of all living organisms, plus even such a simple mechanism as a mousetrap. At another extreme, one might demand the use of written language, or the most complex levels of understanding known to human thinkers—the creative insights of Beethoven, Einstein, or Whitehead, for instance. But these requirements would eliminate many members of our own species.

It is important to recognize at the outset that almost any concept can be quibbled to death by excessive insistence on exact operational definitions. Even such widely used and clearly useful terms as "hunger," "memory," "aggression," or "metabolism" have been subjected to erudite analyses in a search for definitions that will satisfy all demands and avoid every possible ambiguity. These efforts tend to come in waves, each followed by a truce of sheer exhaustion, after which the term continues to be used, but with clearer appreciation of the breadth of its connotations. Excessive concern to avoid all terms that cannot be rigorously defined suffers from the danger of retaining only verbal corpses that display rigor mortis.

It is appropriate to begin with the most obvious fact about mental experiences: *all of us have them*. Every normal person thinks about objects and events. These may be close at hand in time and space, like a toothache, a frightening antagonist, or a beloved infant. But we can also think about things that are remote from the current local situation; our thoughts may concern some distant place and they can reach far into the past or future. Some mental experiences are as simple as recalling the color of last summer's flowers or yearning for satisfying food; others are elaborate and complex, like an astronomer's concept of stellar evolution. I suggest that, for present purposes, we consider an animal to have a *mind* if it has such experiences, whether they be simple or complex.

Awareness involves the experiencing of interrelated mental images. This may not be a completely sufficient definition of conscious awareness, but it certainly captures one of its important properties, one which promises to be particularly useful as we inquire whether conscious awareness also occurs in other species. As discussed in Chapter 2, mental images may be static, like instantaneous sensations, but more often they change continuously as one thinks about the flow of events. If these events include participation by oneself, we say the organism is self-aware. An important class of mental experiences are *intentions*, in which the intender pictures himself as an active participant in future events and makes choices as to which sort of image he will try to bring to reality.

The distinction between conscious awareness and responsiveness is of the utmost importance, and a moment's reflection reminds us that the two are not necessarily associated; at least in ourselves either can occur without the other. It is important also to consider what we know from our own experience about the changing degree of conscious awareness which accompanies a given sort of behavior under different circumstances. For instance, when we are in the process of learning some difficult and moderately complex behavior, we are likely to be conscious of it in much greater detail than we are after it has become a familiar and well-practiced part of our behavioral repertoire. Motor skills, such as swimming, playing musical instruments, or operating automobiles and airplanes, are clear examples. We think consciously about many details of these activities when we are learning them or have just succeeded in doing so. However, after we have used these skills repeatedly for a long period, we usually think instead about the larger behavior patterns into which they fit. For instance, "I must stop at once or I'll hit that child in the road" rather than "I will now flex my right leg and then extend it vigorously against the brake pedal."

We can think consciously about even such basic patterns as breathing, if we make the effort to do so. Much of our difficult and rewarding learning and teaching serves to develop explicit awareness of activities which the student previously carried out with no conscious understanding. For example, athletic skills are often improved by careful consideration of such details as how to hold a golf club or how to breathe while swimming. Once the improved technique has been mastered it, too, ceases to involve conscious awareness. Complex behavior regulated with great precision by our central nervous systems is very often, perhaps in the great majority of cases, conducted without any involvement of conscious awareness. Yet we at least have the capability of, so to speak, turning conscious awareness on or off with respect to particular behavior patterns. It seems quite reasonable to start the consideration of mental experiences in other species by referring to those about which we know at least a little at first hand. Recognizing that mental experiences of any other species may be quite different from our own does not mean that they are nonexistent.

When we turn to other species, we can make at least limited use of analogies to our own situation, although this must be done cautiously, with full recognition that they may be misleading and with every effort to obtain independent evidence against which to check the results of this particular type of reasoning by analogy. In some quarters, reasoning by analogy is disparaged, but on close examination this disparagement turns out to be highly selective; often it is a tactical argument to downgrade whatever type of reasoning is unpopular with the particular advocate. If instead of "reasoning by analogy" one employs a trendy expression like "construction of models," the process takes on an aura of sophistication and a comfortable respectability. But we should ask ourselves whether there is any fundamental distinction between an analogy and a model. Proceeding along these lines, we may infer that it is somewhat more likely than not that when other animals are nearing the end of a difficult learning process and have just mastered some new and rewarding skill, they are likely to be consciously aware of what they are doing. Conversely, we may suppose that when this learned behavior has been repeated for hundreds of times and the behavior has become relatively constant, the animal's attention may well wander and it may think about other things or nothing at all.

Mental images obviously vary widely in the fidelity with which they represent the surrounding universe, but they must exist in some form in any conscious organism. One can scarcely be aware in a general vacuous sense; one must ordinarily be aware *of* something. (Here *thing* obviously includes events and processes, as well as concrete objects.) The presence of mental images, of which the organism is indeed aware, and their potential use by that organism to regulate its behavior, thus provide a pragmatic, working definition of one important kind of *consciousness* which might sometimes occur in nonhuman animals. There are, of course, many other sorts of human consciousness, but it seems tactically sensible to inquire first to what extent such elementary forms of awareness are found in nonhuman animals. Twentieth-century psychology and psychiatry have provided abundant and convincing evidence that we are aware of only a small fraction of the information that flows here and there within our brains, is stored, retrieved, edited, and reformulated into new patterns that affect behavior. But the importance of the Unconscious should not lead us to overlook or underestimate the obvious fact that we are also consciously aware of some mental experiences, and that these are often of great significance. Who would wish to live as a sleepwalker from birth to death? We do, on the whole, enjoy being consciously aware, and it is difficult to imagine human affairs without any conscious component whatever. Granting that this component is important to us, it seems reasonable to suppose that, insofar as it exists, consciousness may also be significant to other species.

Another type of mental experience that may occur in nonhuman animals is belief that something has happened or will happen in the future. Beliefs often entail a sort of propositional relationship between mental images. Human words and sentences can at best be only rough approximations of any actual thought processes that might occur in animal brains, but imperfections of the postulated translation should not blind us to the likelihood that animals might experience thoughts roughly described in the following words: A hungry wolf, "If I chase that deer, I can catch it, and it will taste good." A ground squirrel, "If I dig this burrow deeper, I can crawl into a dark hiding place." A cottontail rabbit, "If I run into this briar patch, that big, threatening animal won't catch and hurt me." Or a male songbird, "If I sing loudly enough, that other male will leave my territory." Such plausible examples can easily be multiplied by anyone familiar with animal behavior.

The learned behavior of laboratory animals can be interpreted in comparable terms, such as (for a white rat), "If I press this lever, food pellets will fall out of that hole in the wall" or (for a pigeon), "If I peck when the light is red, there will be a loud clank, and I can reach some grain." Animals often behave in ways that are consistent with the interpretation that they are thinking in such "if . . . then" terms. Predators stalk, pursue, capture, and eat their prey; rodents dig burrows and take shelter in them; laboratory animals learn new patterns of behavior that yield food, water, or other things they need and use.

It is important to recognize at this point that all of these terms and the concepts they represent are widely and strongly held by many behavioral scientists to be useless for scientific analysis (Lashley, 1923; Boring, 1963; Hebb, 1974, 1978; Skinner, 1974), as discussed in Chapter 8. Because I see some hope of gathering scientifically verifiable data about mental images, intentions, beliefs, and awareness in nonhuman animals, I am naturally troubled that so many of my colleagues have objected to the use of these terms or concepts. One reason for this rejection is the unreliability of introspective verbal reports from human subjects concerning their mental experiences. Another is the wide range of human mental experiences that intuitively seem to involve varying degrees and kinds of consciousness. (For a stimulating discussion, see "Smith," 1975). The primary objection is based on the claim that these terms are not susceptible to precise definition, and hence that their use is detrimental to clarity of thinking. But the best remedy for vagueness is improved evidence that permits clearer understanding. And the search for superior data is impeded, rather than facilitated, by refusal to consider the problem at all. These thorny questions will be examined from several different viewpoints in later chapters.

As a first step toward answering such objections, can we sharpen up these rough-and-ready working definitions sufficiently to permit scientific investigation of animal awareness

without excessive confusion and misunderstanding? One start-
ing point is the accepted usage of scholars working in other
fields. The Random House *Dictionary of the English Language*
defines "awareness" as "having knowledge, being conscious or
cognizant, informed, alert"; "consciousness" as "awareness of
one's own existence, sensations, thoughts, surroundings, etc.";
and "mind" as "the element, part, substance, or process that
reasons, thinks, feels, wills, perceives, judges, etc." We can
also turn to the philosophers, who have been deeply concerned
with the nature of minds. Edwards, when introducing the
mind–body problem in a textbook of philosophy (Edwards and
Pap, 1973), defines "mind" as follows: "Feelings, sensations,
dreams, and thoughts are the sort of phenomena which are usually classified as 'mental'. In calling them mental, philosophers
usually mean that, unlike physical objects, they are 'private' or
directly knowable by one person only." If taken too literally,
this definition would seem to preclude objective investigation of
mental phenomena by anyone other than the person, or perhaps
animal, to whom they are "private." But indirect knowledge is
generally recognized as significant, and we make considerable
use of indirect indications about the mental experiences of other
people. The central question at issue here is to what extent we
can do the same with other species.

Schaffer (1975) defines "mind" as follows: "as the term is
used more technically . . . and in the philosophy of mind today,
[it] encompasses sense perception, feeling and emotion, traits
of character and personality, and the volitional aspects of human
life; as well as the more narrowly intellectual phenomena." Elsewhere Schaffer states: "One thing that sharply distinguishes
man from the rest of nature is his highly developed capacity for
thought, feeling, and deliberate action. Here and there in other
animals, rudiments, approximations, and limited elements of
this capacity may occasionally be found; but the full-blown development that is called a mind is unmatched elsewhere in nature." A cognitive ethologist may wonder whether perhaps the
mental capabilities of animals will turn out to be more substan-

tial and significant than Schaffer implies. To define mental experiences as uniquely human certainly discourages inquiry into the possibility of their occurrence in other species, and begs the question I am trying to examine.

Kenny, Longuet-Higgins, Lucas, and Waddington (1972) devoted a lucid, thoughtful, and stimulating series of Gifford Lectures at Edinburgh to *The Nature of Mind*, without explicitly defining the terms "mind" or "mental." Ryle (1949), in a very influential book entitled *The Concept of Mind*, also avoided any specific definition of the term. But in a second series of Gifford Lectures (Kenny et al., 1973, p. 47) Kenny stated that "to have a mind is to have the capacity to acquire the ability to operate with symbols in such a way that it is one's own activity that makes them symbols and confers meaning on them." The communicative dances of honeybees certainly satisfy this criterion; for it is each forager's own activity that makes the waggle dance into a symbolic statement that conveys to other bees useful information about distance, direction, and desirability of something the dancer has visited. (This communication system of bees is lucidly described by Frisch, 1967; Lindauer, 1971a; and Hölldobler, 1977; and I will analyze it in more detail in Chapter 3.)

Some philosophers may object to calling the bee dances symbolic on the ground that only thinking creatures can recognize symbols, so that use of the term *symbol* implies that bees do think, and thus tricks the reader into accepting the conclusion at issue. For the moment, I mean to point out simply that the bee dances satisfy the particular definition advanced by Kenny. In Chapters 3 and 4, I will discuss in greater detail how this animal communication system provides at least suggestive evidence of conscious thinking. Elsewhere in these Gifford Lectures, Longuet-Higgins offered quite a different sort of definition: "An organism which can have intentions I think is one which could be said to possess a mind [provided it has] . . . the ability to form a plan, and make a decision—to adopt the plan" (Kenny et al., 1972, p. 136). Many animals behave as though they do have plans of at least a simple sort, and adjust their behavior appropriately in attempts to carry them out.

The neurophysiologist E. R. John (in Thatcher and John, 1977, pp. 294–304) defines consciousness as "a process in which information about multiple individual modalities of sensation and perception is combined into a unified multidimensional representation of the state of the system and its environment, and integrated with information about memories and the needs of the organism, generating emotional reactions and programs of behavior to adjust the organism to its environment. . . . Consciousness about an experience is defined as information about the information in the system, that is, consciousness itself is a representational system. . . . Perhaps our philosophical quandary [concerning mind-brain dualism] arises from the assumption that organized processes in human brains are *qualitatively* different from organized processes in other nervous systems or even in simpler forms of matter. Perhaps the difference is only quantitative; perhaps we are actually not as unique as we have assumed."

Conscious awareness and mental experience may sometimes be limited to a single sensory modality, for example vision, so that a rigid requirement that consciousness entail integration across modalities may not be justified. It might also improve John's definition to add that conscious minds deal with dynamic mental images of future, as well as past, events. But in other respects this definition is close to the cautious and tentative views of many neuroscientists, and it is important to note that it does not limit conscious awareness to our species.

2

Do Animals Have
Mental Experiences?

Even though philosophers have not settled on a single explicit and generally acceptable definition of mind, it would be foolishly inhibiting to give up all attempts to understand whatever processes give rise to such concepts as awareness or intention until they can be defined with the assurance with which a chemist assigns a structural formula to a purified compound. When the nature of a sample is unknown, it would inhibit the chemist needlessly to insist that he consider only one of a set of well-defined molecules. The sample might contain something never before analyzed, or, more likely, some mixture of known and unknown substances. A comparable situation may confront us with regard to mental experiences. What now seems to be a single, though vaguely defined, entity such as awareness may well turn out, when fully understood, to be a mixture of known or unknown processes. But we will make little progress if we throw up our hands in dismay and refuse to study the unknown sample at all.

Mental Images

When we think about objects and events, our awareness must be based on representations in our brains of some aspect of the world around us that includes both spatial and temporal relationships. I am using the term *image* in this broad sense to include any representation of the outside world of which persons or animals may be aware—that is, that they may think about and manipulate internally. It need not be a visual image, but may be a pattern of remembered or imagined sounds, smells, or tactile perceptions. Although Ryle (1949) argued that talking and thinking about mental images is foolishly confused, other philosophers, such as Fodor (1968) and Haynes (1976), find them useful concepts. Dennett (1978) tries in an entertaining fashion to explain away mental images, and although I find his arguments unconvincing, his lucidly penetrating analysis is constructive and stimulating.

A mental image ordinarily resembles a sensation or perception, except that it is not necessarily linked to current sensory input. Like perceptions, mental images are not static; they change with time, and one of their important properties is their temporal organization—that is, the pattern of their sequence, duration, and stability or change. Mental images or representations, experienced when an organism thinks about objects and events, seem to offer one category of mental experience which may be more easily accessible than others to scientific investigation. If one is uncomfortable with the term "mental," the term "internal representation" is almost equally appropriate, provided one recognizes that these are often dynamic and adaptable according to the animal's situation and needs. Although once disparaged as meaningless by strict behaviorists, human mental imagery is receiving renewed scientific attention, as reviewed by Segal (1971), Sheehan (1972), Nicholas (1977), Posner (1978), Shepard (1978), Kosslyn et al. (1979), and Kosslyn (1980). A scientific journal called *Mental Imagery* has recently been established. To be sure, the existence of a technical journal does not rigorously demonstrate the reality of its subject matter (witness parapsychology, for example), but it does testify to a substantial interest.

Mental representations of the outside world need not be strictly iconic—literal pictures in the brain—and they may be coded in any of a variety of ways. A simple and well-known example of nonpictorial representation is the nearly universal coding of stimulus intensity in terms of frequency of nerve impulses. Size of a salient object or loudness of a startling noise do not generate large or loud nerve impulses; yet the magnitude of something outside the animal is effectively represented, at least temporarily, within the brain. How magnitude is coded over long periods in stored memories is totally unknown, because we have only vague and contradictory notions about the basic nature of memory traces themselves.

On the other hand, neurophysiologists have known for decades that certain simple aspects of sensory information actually do activate parts of the brain in spatially ordered patterns, thus establishing at least a limited sort of geometric representation. Two clear examples are the reproduction (with some distortion) of the retinal geometry of images in the visual cortex and the tonotopic representation of different frequencies at several levels in the vertebrate auditory systems. Spatial representation of important aspects of the outside world has recently been shown to extend beyond this simple level. For example, Knudsen and Konishi (1978) have shown that the auditory system of the barn owl includes areas of the midbrain in which cells responding to sounds that arrive from different directions are arranged in such a way as to map surrounding space. This "map" is not merely a reconstruction of patterns on a sensory surface, such as the retina or cochlea; it results from neural processing of information that reaches the animal in a wholly different form, such as time or intensity differences at the two ears.

In a particular species of echolocating bats, Suga and O'Neill (1979, 1980) have uncovered an example with even more profound implications by showing that several properties of acoustic echoes are mapped systematically on the surface of the auditory cortex. These properties include not only frequency but intensity, direction of incidence, and distance to the object from which echoes are returned. Thus, neurophysiologists have found at least simple maplike arrangements of responsive cells in both owls and bats. All those so far discovered have involved aspects of the surrounding world for which the animal in question has highly specialized neurophysiological mechanisms for processing sensory information. It may be that the brains of other species contain nothing of the kind or that no other features of the outside world are mapped so iconically even within the elegant brains of owls or bats. But it would be rash to conclude that the neurophysiology of sensation and perception is fully understood.

Clearly, the brain of a perceiving organism need not contain such geometrical analogues, recognizable from neurophysiological data. Yet, in one way or another, every perception, every memory, and every anticipation of future events must be causally related to some representational process within the central nervous system. Of course, a maplike sensory projection, even one constructed from highly noniconic sensory inputs, does not demonstrate that the animal in question is consciously aware of the patterns thus represented in its brain—even though they are crucial to guiding its behavior. But both the behavior and the data about brain mechanisms are certainly suggestive evidence that the animal may sometimes be aware of these patterns. In later chapters I will explore how further evidence might be obtained.

Psychoneural Relationships

Most behavioral scientists, psychologists and ethologists alike, are thoroughgoing materialists. They believe, or at least operate on the working hypothesis, that the processes leading to what we call mental states depend directly upon complex activities of central nervous systems, especially interactions between various excitation patterns. To suppose that mental experiences occur in animals requires no more recourse to vitalism or immaterial entities than does their recognition in ourselves. Some small fraction of the activity in human brains generates what we call mental experiences. In the present state of its development, neurophysiology cannot determine whether there are significant qualitative differences between processes that are and are not accompanied by subjective mental experience.

Postulating the existence and significance of mental states in other species implies little or nothing about the functional relationship of such states to the central nervous system. For the purpose of this discussion, it makes little, if any, difference whether one assumes that mental experiences are identical with some pattern of neural activity, or whether one prefers some other type of psychoneural relationship (such as those discussed in the volume edited by Globus et al., 1976). The basic question I am trying to answer is the degree of similarity of mental experiences resulting from brain function in our own and other species. Only if one postulates a major, qualitative difference between psychoneural relationships across species does this complex area of philosophical concern become directly relevant.

Of course, any evidence bearing on the mind-brain relationship is important in its own right, and insofar as it involves observable mechanisms, it becomes pertinent to inquire whether these may, in fact, differ among species. Lateral control of linguistic communication by one side of the cerebral cortex is a clear case in point. Relatively small anatomical differences between the right and left temporal region of the human cortex have been related to this functional asymmetry, which often is considered a unique feature of human brains. This asymmetry has often been advanced as neurobiological evidence of a species-specific human mechanism closely related to speech, and hence by implication to conceptual thought (reviewed by Brown, 1976; Neville, 1976; Galaburda et al., 1978; Gazzaniga, 1975, 1979; Levy, 1979; Nottebohm, 1977, 1979; and in conference proceedings edited by Harnad et al., 1976, and by Dimond and Blizard, 1977).

But the localization of a certain function in one part of the brain does not in itself tell us much about just what is occurring there. Many animal brains are slightly asymmetrical, some more so than the human brain. Especially important are the recent discoveries of Nottebohm (1977, 1979), which have demonstrated that in songbirds the control of vocalization is almost entirely concentrated in one half of the brain. These cerebral asymmetries are much more pronounced than those found in the human brain. Should we therefore accept songbirds along with signing apes into our Select Kingdom of talkers and thinkers? Many otherwise normal people do *not* have speech-control centers that are demonstrably larger in one hemisphere (Galaburda et al., 1978); should they be banished from humanity? If we define human uniqueness on too narrow a foundation, we are in danger of having it undermined whenever the same feature is discovered in some other species.

Ryle (1949) ridiculed the belief he ascribed to many philosophers that a person's mind is a "ghost in the bodily machine." Yet, like Skinner (1957, 1974), he freely recognized that we do think and experience internal representations of external objects and events. The "ghost" that both Ryle and Lashley (1923) were trying to exorcise is derived from belief in an extramaterial realm, a sort of mental universe inherently different in kind from the material, physical world. One can agree completely with Lashley and Ryle about this point and still ask meaningful questions about the thinking or experiencing of internal images in other species. The scientific consideration of animal awareness does not require any of the following: (a) ascribing to other species anything approaching the human level of intellectual capacity, (b) postulating immaterial mental essences, or (c) endowing animals with immortal souls (Humphrey, 1977). One certainly need not depart so far from common sense and everyday experience as physicists have done in postulating antimatter. It seems most reasonable and parsimonious to postulate, tentatively and pending new evidence, that thinking and experiencing are related in comparable ways to the functioning of central nervous systems in various species. It contributes very little to our understanding of these difficult problems to erect and then demolish straw ghosts.

A confident belief in biological evolution leads me to expect that although mental experiences of other species may differ greatly from ours, they will turn out, when fully understood, to share important properties with the entities we meet through our individual introspection. But of course this is a question to be kept open and under investigation; dogmatic assertions are wholly premature. To put the question in slightly different terms: How do those results of brain activity that we call mental experiences differ among various species of animals? The attempt to study animal thinking was largely abandoned early in the twentieth century, but so much has since been learned about animal behavior, and how it can be objectively analyzed, that a cognitive approach to ethology has much better prospects of success than in the days of Darwin and Romanes.

Some psychologists claim that they and their colleagues have been studying mental experiences of animals when analyzing learning, problem-solving, discrimination, and the like (Mason, 1976). But, since the pioneering experiments of Köhler (1925), such subjects have been discussed by experimental psychologists almost exclusively in terms of the *behavior* of the animals concerned, rather than any possible awareness or mental *experience*. Behavioral scientists conclude that an animal has learned to do something it did not do before, but seldom, if ever, do they ask what the animals may be thinking or feeling. We have stopped asking whether they *know* or *believe that*. . . . As will be discussed in Chapter 8, the principal reason for this disinterest in mental experiences of animals has been the behavioristic viewpoint which entails a sort of negative injunction not to be concerned with mental experience in either men or animals. The negativity is even stronger with nonhuman animals than with our own species, and a tacit assumption of human uniqueness tended to creep into the behavioristic *Zeitgeist*. This can be summed up as the assumption that people talk about alleged feelings, beliefs, intentions, and other mental experiences; but animals have none of these experiences, and even if they did, they don't talk and hence could not tell us about them.

It is obvious and important that our mental experiences include not only images and intentions, but also feelings, desires, hopes, fears, and a wide variety of sensations such as pain, hunger, rage, and affection. All such subjective entities have been largely rejected by traditional twentieth-century psychology and ethology for the same basic reason that they are "private data" directly observable only by the one who experiences them and describable to others only by introspective reports (Alston, 1972). The qualities of sensations are especially troublesome and have almost wholly eluded objective definition. A classic example is that no one has been able to propose a satisfactory method to prove or disprove a statement such as "Your sensation of blue, stimulated by light of 450 nanometers, is exactly like what I report as red when my eyes are stimulated by 625 nanometer light." Each of us has learned to identify his private sensation according to the names given to light of these wavelengths or of objects that send such light to our eyes. Hence, agreement about nomenclature tells us nothing about the sensations actually experienced.

In preliminary efforts to come to grips with the question of possible mental experiences in animals, I prefer to concentrate on images, beliefs, intentions, and awareness of objects and relationships in the outside world, rather than on feelings and purely subjective qualities. My reason for this choice is that I can see more realistic hopes of developing objective methods for gathering satisfactory data about the former than about what psychologists have called "raw feels" (Tolman, 1932). In Chapters 8 and 10, however, I shall return briefly to the question of subjective feelings in animals.

According to the working definitions presented above, it is not necessary to assume that consciousness or mental images are present only in living organisms; computers have often been suggested as likely candidates (Scriven, 1963; Apter, 1970; Elithorn and Jones, 1973; Gregg, 1974). Strong dissent from this viewpoint has been expressed, however, by Searle (1980). But the relationship between minds and computers is outside the scope of this book, which will confine itself to the possible existence of mental experiences and awareness in nonhuman animals. It may be helpful, however, to bear in mind the analogies suggested by Longuet-Higgins in Kenny et al. (1972); central nervous systems may be likened to the hardware of computers and minds to the software, or programs. Each of us has his own mental experiences and thus, in a sense, sees some of his software from inside. The strict behaviorist insists on ignoring this source of information, on the ground that it cannot be observed directly by another person. But this may be as great a limitation as to observe a working computer and deny the existence of a program guiding its operation because the program cannot be seen on close inspection of the teletype, recording tape, or central processing unit.

Despite the self-consistent philosophical positions which deny that the nature and reality of mental experiences in other human beings can ever be demonstrated, I suggest that we accept the reality of our own mental experiences, even without rigorous proof. This suggestion runs counter to a strong prevailing tendency to dismiss mental experiences as insignificant, if not unreal. This viewpoint is eloquently set forth by Dennett (1978), among others. A major component of these arguments is based on the positivist view that only objectively verifiable data should be used in constructing theories or forming opinions about everything in the universe, including even the thoughts of human beings and, in particular, those of the scholar himself. It certainly seems far more likely than not that mental experiences, whatever their actual nature may be, are closely linked to neurophysiological processes within our brains, even though we may not yet understand these processes at all well, and many important neural functions remain undiscovered. One possibility is that the relationship between mind and brain has important elements in common with the relationship between properties of whole animals and those of their constituent cells.

A helpful analogy is provided by considering a well-known and noncontroversial class of behavior—coordinated locomotion. No one supposes for a moment that an animal's walking, running, swimming, and flying require more than the activities of muscle cells, connective tissue, and neurons; no immaterial "essence of locomotion" is called for. But the patterns of structural and functional coordination by which thousands of cells produce bird flight, for example, are not easily derived from data on the endoplasmic reticulum or sliding filaments of acto-myosin. If only for practical reasons, we are forced to deal with bird flight and similar examples of coordinated locomotion in terms appropriate for their level of organization; but this does not delude us into postulating vitalistic essences of flight inde-pendent of physics, chemistry, or cell biology. Although many details of locomotor physiology require further clarification, few, if any, scientists doubt that full explanations are possible in physiological terms. Nor need this confidence diminish our ad-miration for the success and beauty of the behavior under study. In some comparable fashion, I suspect, minds depend entirely on the functioning of central nervous systems (including neu-roendocrine mechanisms), yet exhibit properties not easily pre-dictable from even the most complete analysis of neurons and synapses. Similar concepts have been advanced by Lorenz (1963), Scriven (1963), K. Smith (1969), Sperry (1969), Piaget (1971), Eccles (1973, 1974), Popper (1974), and Hubbard (1975), as well as by several contributors to a symposium edited by Ayala and Dobzhansky (1974). But the full impact of these ideas has not yet been recognized by students of animal behavior.

Self-Awareness

An increasing number of behavioral scientists seem willing to consider the possibility that animals are sometimes aware of objects and events, but insist that self-awareness is unique to our species. This is one of a very few areas of cognitive ethology that have already been illuminated by objective, verifiable experiments. Gallup (1970, 1975, 1977) has developed an ingen-ious procedure which exploits the interest displayed by chim-panzees in their own mirror images. These and other Great Apes have often learned to use mirrors to examine parts of their bodies which they cannot see directly (Hayes, 1951; Hayes and Hayes, 1951). Gallup gave four chimpanzees ample opportunity to use mirrors, while two others had no such experience. In the crucial experiments, one animal at a time was deeply anesthetized and a conspicuous spot of inert, quick-drying colored material was placed on its forehead or earlobe. Upon awakening, they paid no attention to the markings, indicating that local, tactile stimulation was absent or ineffective. But when a mirror was provided, the chimpanzees familiar with their mirror images looked at themselves and then almost immediately reached for the colored spot and rubbed it or picked at it with their fingers. Those lacking experience with mirrors continued to ignore the paint marks. Certainly this experiment suggests that they recognized the mirror image as a representation of their own bodies.

So far, Gallup's type of experiment has yielded positive results only with Great Apes. Despite intensive efforts, gibbons, monkeys, and other laboratory animals have failed to react to mirror images as replicas of their own bodies. Instead, they seem to treat the mirror image as though it were another animal. Gallup therefore concludes that no other animal has the capacity for self-awareness. But one can inquire whether the capability of responding appropriately to mirrors should be equated with self-awareness. Would other tests prove more suitable for other species and provide evidence for the concept of oneself? Only further investigation can answer such questions; meanwhile, Gallup's experiments provide a clear and successful example of a well-controlled, objective, verifiable experiment in cognitive ethology. Ingenious experiments by Beninger et al. (1974) shows that rats can learn to respond differently to a signal according to their behavior at the time the signal occurs; this suggests awareness of their own activities and hence, perhaps, of themselves.

It is instructive to separate the possibility that animals may be capable of self-awareness from the larger question of whether they have any awareness at all. If not, the question of self-awareness obviously does not arise. Therefore, let us explore the first possibility—that animals are capable of *some* kinds of awareness, but not of self-awareness. This means that we assume the animal is aware of its companions, their actions and communication signals, as well as of its own physical surroundings, the ground beneath its feet, the wind that blows against its skin, and so forth, but that it is unable to entertain the concept of "selfness." Yet an abundant flow of sensory input is always arriving at the animal's central nervous system from its own body. So we must postulate that this input is somehow selectively barred from reaching the animal's awareness.

This kind of "awareness of everything but me" is conceivable, but it becomes increasingly less plausible the wider the range of awareness the animal is postulated to have of its inanimate and animate surroundings. If we allow a particular animal to be aware of a reasonably wide range of objects, events, and relationships in the world around it, while denying the possibility of self-awareness, we run the danger of redefining self-awareness in a roundabout way as a sort of perceived hole in the universe. Self-awareness has been widely held to be absent from all species but our own (for example, see Popper and Eccles, 1977). Premack (1976) and others assert without qualification that not even Great Apes are aware that they themselves will die. But direct and unequivocal evidence is nonexistent, and this whole subject challenges cognitive ethologists to seek for relevant data.

This Chapter has reviewed many of the principal issues which make it of the utmost importance to clarify the differences between mental experience in human and nonhuman animals: Do mental experiences occur in nonhuman animals? Does complex behavior such as that required for the more impressive examples of animal orientation indicate the presence of conscious thinking? When animals learn some new task or sensory discrimination, are they ever consciously aware of the facts and relationships they have learned? Can scientists investigate animal awareness in a balanced, open-minded fashion without undue reliance on appealing, but unsupported, assertions? Can mental images, consciously perceived, be detected in nonhuman animals, and if so how? Do animals experience beliefs and intentions, and are they aware of themselves?

It is now time to examine one category of animal behavior that is perhaps most closely linked to whatever they may be thinking—their communication. Animals often go to great trouble to direct messages at conspecifics, signals that produce important changes in their behavior, sometimes including exchanges of communicative signals which lead, in turn, to coordinated group action. When such social communication is versatile and appropriately adapted to a variety of situations, the possibility arises that the communicating animals are thinking about the messages they exchange with their fellows.

3

The Versatility of
Social Communication

The analysis of communication between individual animals has led to several discoveries of the highest significance. Investigations of a wide variety of species belonging to several divergent branches of radiating evolutionary lines of descent, from fiddler crabs (Crane, 1975) and Hymenoptera (Hölldobler, 1977) to dolphins (Herman, 1980*), monkeys and apes (Goodall, 1971, 1975; Marler and Tenaza, 1977; Seyfarth et al., 1980 a and b) have demonstrated a common thread of versatile diversity. Although something simpler was initially expected, communication signals have turned out to include, at the very least, an announcement that the sender is of a given species, sex, and appropriate age, and is in one of a relatively few basic behavioral states, such as readiness for fighting, fleeing, or mating (reviewed by Smith, 1977, and in the volume edited by Sebeok, 1977). These messages also have an intensity scale from weak to strong.

Conspecific partners respond to varying degrees and in different ways, but often appropriately according to their own age or reproductive condition. Individual recognition, or at least differential reaction to communicative signals from particular conspecific companions, is common at least in birds and mammals (Beer, 1973a, 1973b, 1975, 1976; Falls, 1969, 1978; Falls and McNicholl, 1979; Green and Marler, 1979; Cheney and Seyfarth, 1980). Communicative behavior is adapted and modulated in accordance with the responses of other animals. A frequent element is the flexibility and interrelatedness of the signaling behavior; fairly complex sequences are performed, with each step depending on the signals or responses returning from the partner. Many species of birds, for example, exchange long sequences of songs or calls, which are modified by the replies of other birds nearby (Kroodsma, 1978). In a few species dueting is common, and mated pairs or fellow members of communal roosting areas may modify their calls to resemble those of their social partners (Thorpe, 1972a; Mundinger, 1970; Nottebohm and Nottebohm, 1976).

Rather than being fixed motor patterns determined solely by the physiological state of the animal or by its physical environment, the essence of communication is interaction with one or more other animals. Interaction involves changing one's own behavior in response to the behavior and signals of another animal. Furthermore, it is often of considerable importance that each communicating animal gauge as accurately as possible the mood and the behavioral dispositions of the others with which it is in communication. To the extent that nonhuman animals may have mental experiences, especially beliefs or intentions, it may also be important for their social companions to understand these mental states. It is therefore possible that an animal's communication provides relatively direct access to any mental experiences that it may have.

Almost every sensory system is employed by some species of animal for communication with conspecifics. Chemical signals, including pheromones, are ordinarily detected by the olfactory system and are especially important in insects and such nocturnal mammals as rodents and bats (Shorey, 1977). Sounds are used extensively by many groups of arthropods, as well as by all classes of vertebrate animals (reviewed in volume edited by Sebeok, 1977). Surface waves are used by aquatic insects (Wilcox, 1972; Markl et al., 1973). Tactile communication includes both direct contact between animals and communication via vibrations of the ground or vegetation. Leaf-cutter ants stridulate when they are buried accidentally, and other members of the colony locate them by vibrations transmitted through the soil (Markl, 1967, 1968, 1970; reviewed by Wilson, 1971).

In certain spiders, the male begins his courtship by setting the female's web into a particular pattern of vibrations. Many groups of fishes that use electrical orientation also communicate by electrical signaling (Hopkins, 1974, 1977, 1980; Westby, 1974, 1975a–c)

Communication by visual signals is widespread (Marler, 1968; Hailman, 1977a, 1977b). An especially striking example is the courtship of certain fireflies, which exchange light flashes signaling sexual readiness (Lloyd, 1977, 1979). The females of some predatory fireflies not only respond to flashing patterns of conspecific males, but at one stage in their life cycle they mimic the courtship patterns flashed by receptive females of smaller species. This commonly lures courting males of the latter species to their death at the mandibles of these firefly "femmes fatales" (Lloyd, 1975, 1980). But visual signaling has not been studied as extensively as has acoustical communication, primarily because it is technically more difficult to record and play back most types of visual signals.

Granted that animal communication is more complex, varied, and versatile than we used to suppose, does this tell us anything about the occurrence of actual thinking or feeling on the part of these communicative creatures? We have good reason to associate talking with human thinking, and human feeling with nonverbal communication. In later chapters I will discuss the possibility that comparable inferences from animal communication to animal experiences may be warranted. To be sure, many behavioral scientists believe that such an approach is futile, misguided, or both. One reason for this skepticism is the conviction that when animals communicate they are always reacting only to the immediate situation and emit only signals closely linked with their emotional states. Never, it is claimed, do "thoughtless brutes" convey information about objects or events that are not immediately present. Holding this opinion makes it easier to persuade oneself that animals are unthinking and unfeeling automata.

A helpful first step in considering this large question is to examine a few specific examples of nonhuman animals that exchange messages of moderate complexity and adapt such communicative behavior to changing circumstances. I will select for special consideration several cases in which there is some evidence that animal communication is not limited to the immediate flux of impinging stimuli. This is another way of saying that it seems especially pertinent to analyze cases where animal communication signals have the property often called *displacement*, meaning that the signals refer to, and convey information about, objects and events that are removed in space or time from the immediate situation. In short, I will inquire whether animals communicate, and even perhaps think, about memories of past events, about anything that is not right here or occurring right now, or about what may happen in the near or distant future.

When an animal reacts to immediate stimuli it need not do so mechanically without any awareness. Conversely, a person or animal may not be conscious of his reactions to distant events, or even his communication about them. Intuitively, it seems more likely that communication which has the property of displacement will be accompanied by awareness corresponding at least approximately to the content of the transmitted message. This intuitive feeling certainly seems generally correct for human language, although exceptions can be found. That is, when we talk about something not immediately present, and are correctly understood, both speaker and listener almost always think about at least some aspects of the message thus conveyed. Whether animal awareness (insofar as it occurs at all) bears a similar relation to displacement remains an open question.

Gestural Communication with Displacement

Solitary creatures need not communicate, and it is self-evident that human language is widely used to affect the behavior of our companions. A basic requirement for social cooperation is some sort of communication to coordinate group activities, whether in hunting larger or more elusive prey than a single predator could catch on his own, or accomplishing any other mutually beneficial task by joint effort. Animals that live in mutually interdependent social groups may thus be more likely to engage in communicative behavior from which we might learn something about what, if anything, they are thinking. It is also helpful to inquire whether social animals ever have a serious need to communicate about objects and events that are not immediately present. Do vulnerable species ever warn each other to avoid places where one of them knows a dangerous predator is waiting? Do they ever exchange such messages as: "Don't go to the water hole, I saw a leopard there?" No convincing evidence of communication with this degree of specificity has been reported by ethologists, although frightened animals often convey to their companions at least generalized states of alarm and wariness which serve much the same function. Familiar examples are the alarm calls of chipmunks or the "blowing" of startled deer. Parents commonly lead or guide their dependent young from one place to another, and in so doing they generally select safer routes and go to more favorable locations while avoiding sources of danger and excessive exposure to the elements, poor food supplies, and the like.

Food is all-important and, under natural conditions, it may be available only briefly at widely separated and unpredictable locations. Especially when animals have dependent young, they must often search intensively and exhaustingly for food which they bring back to den or nest to feed their offspring. Food usually seems to be sought and gathered quite independently by the parents without any evidence of cooperation—not even, as far as we know, at the simple level of informing each other that a particular location has just been examined and found to offer nothing edible. If such communication does occur, it would be very difficult to detect. But it would be highly adaptive, in evolutionary terms, for animals to share such information with mates, offspring, and close relatives.

Does one parent locate an abundant source of food and then inform his or her mate where to go for more? In considering such possibilities, it is natural to think of birds with large clutches of nestlings. In many species, both mother and father work long and hard to bring back enough food, and often it would be advantageous to share information about newly discovered food sources. But ethologists have not reported evidence that they do so. It is difficult to determine with confidence whether this is because such communication never occurs, or whether practical difficulties of observation have prevented its discovery.

Some birds produce characteristic calls on finding good food sources (Green and Marler, 1979), and these calls seem to attract others of their species, but no ethologist seems to have observed selective communication of such information to the mate or offspring, as one would expect on the basis of kin selection (Wilson, 1975). Again, it is not clear whether, if such behavior occurs, it would have been detected. It seems likely, though even this has not been well documented, that one parent, say the male, may follow his mate when she returns to a newly discovered food source; this possibility should be testable by appropriate field studies. It has been suggested that an important adaptive advantage to living or nesting in large colonies lies in the sharing of information about changing food sources (Ward and Zahavi, 1973; Krebs, 1979). However, little attention has yet been paid to whatever communicative behavior may serve to transfer such information from one colony member to another. One might assume that colonial birds observe the direction from which others return with food, but very few specific observational data are available (Krebs, 1974, 1979).

It is often necessary that one food-gathering parent stay at the nest to incubate the eggs. Could this bird in some way instruct its mate where to find food? Might a mother bird convey to the father any message at all comparable to: "While I clean feces out of the nest, you go fetch more of these juicy caterpillars from the swamp over beyond the oak woods?" This notion seems so ridiculously anthropomorphic that few, if any, ethologists would entertain it even as a hypothesis to be tested by appropriate observations. Birds and nonhuman mammals do not seem to be capable of communication with this degree of displacement.

Perhaps it would be profitable to inquire whether any other animals live in more highly interdependent social groups and have problems of food supply that would make it even more advantageous to convey such information to one another. Are there any social primates which must forage far and wide over dangerous terrain for scarce and unpredictable food supplies? If so, might certain individuals specialize, at least for a time, on searching for food and bringing it home to feed the young, the pregnant females, or other group members which for some reason do not forage for themselves? If this does occur, it would be advantageous for a successful food-gatherer to inform his social companions (which are usually close relatives) where the food supply has just been found. A diligent psychologist could almost certainly instill such behavior in a group of monkeys by intensive training. And it would certainly be adaptive under many natural conditions, so that selection should favor monkeys capable of such cooperative communication. But does anything of this sort actually occur in nature? Not according to most experienced students of primates, birds, or other vertebrates.

The most highly integrated animal societies, with the greatest division of labor and specialization of function among group members, are not those of vertebrate animals. They are found among the social insects—termites, ants, and bees—some of which live in colonies of thousands or even millions. Usually all colony members are sisters or half-sisters, and the food needed for such enormous numbers of adults and young is often gathered by colony members that specialize in locating distant and unpredictable sources which must be exploited as soon as possible before they disappear. Therefore, an evolutionary biologist might well expect that in such mutually interdependent societies, if anywhere, we would find animals communicating about objects remote from the immediate situation where the communication occurs (Humphrey, 1979).

Despite the plausibility of what most zoologists will recognize as "expectation by hindsight," we were incredulous when just this sort of flexible communication with a high degree of displacement was actually discovered through the brilliantly pioneering experiments and insights of Karl von Frisch (1923, 1946, 1967, 1972, 1974). I am, of course, referring to the communicative dances of honeybees, mentioned in Chapter 1. These take several forms; the most significant is the *Schwänzeltanz* (usually translated "waggle dance"). This is a figure-eight-shaped pattern ordinarily carried out inside a hive in total darkness by bees crawling rapidly about over the vertical surface of the honeycomb. The dancers are workers or nonreproductive females—ordinarily the older workers that have been engaged previously in many other complex activities inside the hive. Along with the dances there is a great deal of other communicative behavior involving exchange of tactile and chemical signals. Food materials of several kinds and associated scents are transferred from one worker to another through behavior patterns that so far have been described in only the most general way, but often involve regurgitation of stomach contents and exchange of food materials (Wilson, 1971).

Bees execute these waggle dances most commonly when a forager from a colony that is suffering from a shortage of food has returned from a rich food source and carries either nectar from flowers in her stomach or pollen grains packed into basketlike spaces formed between specialized hairs on her legs. One cycle of the waggle dance consists of a circle with a diameter about three times the length of a bee, followed by a straight portion and then another circle turning in the opposite direction from the first, after which the straight segment is repeated. The circling thus alternates clockwise and counterclockwise. The straight portion is the important component for transferring information, and it is during this part of the figure-eight pattern that the abdomen is moved vigorously from side to side at 13 to 15 times per second. Other bees cluster around the dancer and follow her through the pattern traced by her dance.

Although these waggle dances had been observed for hundreds of years, it was Frisch who first deciphered the messages they convey. This discovery has much in common with Young and Champollion's use of the famous Rosetta stone to decipher Egyptian hieroglyphics, Skinner (1974) to the contrary notwithstanding. For the first time, a simple but truly symbolic communication system of another species was identified and at least partially translated.

Frisch made this revolutionary discovery while studying bees in an observation hive—a small, thin structure with space for only a single layer of honeycomb, but provided with a glass window on one or both sides so that the bees can be seen while carrying out all the many activities involved in maintenance of the hive and care of the queen, eggs, and developing larvae. He was feeding his bees concentrated sugar solutions in small dishes and, for convenience, these dishes were placed close to the hive. The bees were marked individually with small daubs of paint while they filled their stomachs at the feeding stations. When these bees returned to the observation hive they danced by moving only in circles—called round dances—alternately clockwise and counterclockwise, but other bees from the same hive were bringing pollen back from distant flowers. The first new discovery was that the pollen-carriers executed waggle dances, and Frisch (1923) concluded that the two patterns were somehow correlated with the type of food material. Not until more than 20 years later did he happen to study the dances performed by bees that had brought back the same type of food from different distances. He found that the waggle dances then occurred when any food (nectar, artificial sugar solutions, or pollen) was brought from more than roughly 100 meters, showing that the primary factor was distance, rather than type of food. The transition from round to waggle dances is a gradual one, with geometrically intermediate forms, and different varieties of the honeybee *Apis mellifera* and related species begin the shift from round to waggle dances at distances ranging from about three to fifty meters.

Frisch followed up this observation with an extensive series of experiments which demonstrated that whenever bees perform waggle dances after gathering food at distances beyond about 50 to 100 meters, the straight waggling portion of the dance varies in length and direction, and that these variations are closely correlated with the location of the food source. Direction is indicated relative to the vertical and to the location of the food relative to the position of the sun in the sky. Thus, if food is directly toward the sun, the straight waggling run is oriented straight up; if directly away from the sun, straight down; and correspondingly at intermediate directions. The length and duration of the waggling run vary according to the distance from hive to food. Hence, a human observer who knows the code discovered by Frisch can observe the waggle dances and read them well enough to determine the location of the food with an accuracy of approximately ± 5 or 10 degrees and ± 10 percent in distance. A series of these dances specifies quantitatively the distance and direction, and qualitatively the desirability, of what a scout bee has located (the sugar concentration of nectar, for instance). Although directions are ordinarily expressed relative to the vertical, under special conditions honeybees carry out the same type of dance on a horizontal surface in view of the sky, and the waggle run then points directly toward the food.

Desirability is conveyed by the vigor of the dances which can easily be judged by an experienced human observer. The exact properties of dances that constitute their vigor have not been clearly identified, but they seem to include the amplitude of the lateral movements of the abdomen and the intensity of mechanical vibrations that can be picked up by a small microphone held close to the dancing bee (Esch, 1961; Wenner, 1962). If a bee is bringing back a very rich sugar solution when the colony is in need of carbohydrates, she dances a large fraction of the time, and each bout of dancing is repeated for many cycles. If the nectar or artificial sugar solution is very dilute, she may dance only a small fraction of the time, and each bout may include only one or two cycles. More vigorous dances are much more effective in attracting followers and recruiting other workers to fly to the location from which the dancer has just returned.

A dancing bee usually conveys information as to what she is dancing about by transferring odors to other bees which cluster around her. Information from the waggle dance suffices only to bring recruits to the general area of the food source; its exact location must be found by searching for the scent conveyed during the dance. Odors from flowers or scented secretions by the bees themselves add specific information to the dance pattern, and such odors enable the bees to locate the exact position of the food or other desiderata. Sometimes the transfer of odors results from direct contact between odorous particles adhering to the external surface of the dancer and the antennae of other bees. Odors are also transferred along with the stomach contents that are regurgitated by the returning forager and sucked up eagerly by the other bees, which pay close attention to the dancer. Often odors alone are sufficient to enable new recruits to find the food source; and independent searching by individual foragers seems to be adequate under many conditions.

Although the dances are most frequently used to signal the location of a food source, they are also applied to other requirements of the mutually interdependent colony of bees under special conditions. They are called into play primarily, if not exclusively, when the colony of bees is in great need of something. Ordinarily this need is for food, but the communication system is not tightly linked to any one requirement. Lindauer (1955) showed that the same dances are used for such different things as food, water, and resinous materials from plants (propolis).

Human observers can see the patterns of honeybee dances through the window of an observation hive, but the bees do most of their dancing in darkness, because the entrance to an ordinary hive is very small and virtually no light can penetrate into the labyrinthine recesses between multiple layers of honeycomb. Tactile sense organs (mainly those of the antennae and in the joints of the exoskeleton) must be the primary channels by which the dancer transmits information through her waggle run to follower bees, which can somehow relate these complex jostlings to the force of gravity. Later, the follower must also be able to orient her flight at the same angle to the sun as did the dancer to gravity during the waggle dance.

An important aspect of this communicative behavior is its flexibility. When food is plentiful, dancing is rare or nonexistent for long periods. It is quite possible that, under favorable conditions, dancing is not needed and never occurs during the entire life span of the workers. But this point is technically difficult to establish with certainty, because it would be extremely laborious to watch each of several thousand worker bees continuously, 24 hours a day, for a period of weeks. Furthermore, the dances are not fixed patterns performed mechanically and uniformly immediately on return to the hive. Ordinarily they are preceded by extensive tactile contact with other workers, and are part of a complex matrix of social interactions that include exchange of scents and pheromones.

When a colony of bees is engaged in swarming, scouts search for cavities suitable as a future home for the entire colony and report the locations by the same dances, which are now performed by crawling over the mass of bees that makes up the swarm cluster (reviewed by Frisch, 1967, and Lindauer, 1971a). The dances during swarming are especially significant because it seems that no material substance is brought back from the place described by the dancers. However, bees may mark suitable cavities by some appropriate chemical signal, such as Nassanov gland secretions, and use this odor in place of the odors of flowers to convey the information that something desirable was located where indicated by the dance. Furthermore, as discussed in more detail below, the search for, and communication about, cavities is a wholly new experience for the worker bee.

Lindauer discovered another fact of the utmost importance when studying the dances of scouts after they returned to a swarm that had left its original hive. The same marked bee would sometimes change her dance pattern from that indicating the location of a moderately suitable cavity for a new hive to one signaling a better potential site. This occurred after the dancer had received information from another bee, and had flown out to inspect the superior cavity. Thus, the same worker bee can be both a transmitter and receiver of information within a short period of time and, despite her motivation to dance about one location, she can also be influenced by the similar, but more intense, communication of another dancer. Despite the profound implications of these investigations, ethologists have not followed them up with more detailed analyses. The only exception is a careful study by Seeley (1977) that shows how scout bees evaluate the suitability of cavities by exploring them thoroughly and gathering information about their size and other important properties.

There is no escape from the conclusion that, in the special situation when swarming bees are in serious need of a new location in which the colony can continue its existence, the bees exchange information about the location and suitability of potential hive location. Individual worker bees are swayed by this information to the extent that, after inspection of alternate locations, they change their preference and dance for the superior place rather than for the one they first discovered. Only after many hours of such exchanges of information, involving dozens of bees, and only when the dances of virtually all the scouts indicate the same hive site, does the swarm as a whole fly off to it (Lindauer, 1971a). This consensus results from communicative interactions between individual bees that alternately "speak" and "listen." But this impressive analogy to human linguistic exchanges is not even mentioned by most behavioral scientists, for instance Brown (1975), who devotes a whole chapter in his excellent textbook to the dances of bees. We are so accustomed to thinking of insects as genetically programmed robots that it is remarkably difficult to accept the versatility of this communication system at its face value, as discussed in Chapter 5.

Honeybees carry out several other types of dancelike motions, some of which appear, from very limited evidence, to have a communicative function (Frisch, 1967, pp. 278–284). One of these is the *Schwirrlauf*, or buzzing run, performed on the surface of a swarm after the dancing scouts have reached a nearly complete consensus about a particular location. It seems to convey a sort of imperative "Let's go!" and is followed by a mass flight of the entire swarm. Details of the waggle dances might also convey information other than the geometrical patterns related to direction and distance. More information may be conveyed by these less conspicuous motions, and hence it would be premature to conclude that we understand completely the communication behavior of honeybees.

Having received the facts of the situation, it is appropriate to turn briefly to some questions of semantics. Many will object to calling bee dances symbolic, although they meet one of the meanings of "symbol" in the Random House *Dictionary of the English Language:* "Something used for or regarded as representing something else," provided that we may consider gestures and motions to be *things* in the sense of this definition. The waggle dances also satisfy Kenny's criteria for the presence of a mind, as discussed in Chapter 1.

Charles Morris (1946*) proposed a special meaning for the term symbol, one that is applicable to bee dances. He first defined a sign as "something that directs behavior with respect to something that is not at the moment a stimulus," and proposed that a symbol be defined as "a sign produced by its interpreter which acts as a substitute for some other sign with which it is synonymous; all signs not symbols are signals." In the case of bees ready to seek food, the waggle dance substitutes for direct leading of recruits to the food or pointing directly to it. Direct pointing does occasionally occur when honeybees *(Apis mellifera)* dance on horizontal surfaces, and this sort of dance is the only type employed by the dwarf honeybee, *A. florea* (Lindauer, 1971a). Such directly pointing dances might be considered signs in Morris's terms, whereas dances transposed to gravity as a directional reference would qualify as symbols. Even the dances on horizontal surfaces fulfill Morris's definition of a sign about another sign, specifically about the process of actually leading the way to the food. Thus they, too, might qualify as symbols. On this interpretation, the transposed dances carried out on vertical surfaces in the dark hive would qualify doubly as symbols.

Regardless of the semantic issues discussed above, it is clear that the versatility of honeybee dances raises basic questions for which we had been poorly prepared by the behavioristic tradition in psychology or the comparable reductionism in biology. Such complex signaling would have been surprising enough in mammals. Since insect brains weighing only a few milligrams can manage flexible two-way communication, the possibility clearly arises that languagelike behavior may occur in other animals, as well. In other words, the occurrence of symbolic communication in two groups as distantly related as Hymenoptera and Primates (whose evolutionary lines of descent diverged at least 500 million years ago) suggests that such behavior is not the exclusive prerogative of any one species.

Skepticism about Symbolic Communication by Honeybees

The discovery that insects employ a communication system that edges so close to human speech in its symbolism and flexibility has far-reaching implications which may well have played a part in the skepticism expressed by Adrian Wenner (1971, 1974), Wells (1973*), Wells and Wenner (1973), and doubtless felt by many others, such as Hinde (1970), Langer (1972*), Tavolga (1974), or Glucksberg and Danks (1975*). Wenner and his colleagues seriously questioned whether the evidence presented by Frisch and Lindauer really does suffice to demonstrate communication of information about distance and direction. They contended that site-specific odors can account for the results of Frisch's experiments, and that bees do not convey to one another information about distance and direction. They conceded that the dance patterns are closely correlated with the distance and direction of a food source from which the dancer has just returned, but interpreted this as a sort of accidental epiphenomenon. As pointed out by J. L. Gould (1976), correlation between a behavior pattern and some other process does not prove that the behavior serves for communication. But this line of skeptical thought leaves one in the awkward position of having no explanation to offer for the remarkably specific correspondence of dance pattern with the appropriate distance and direction, or for the close attention paid to the dances by other bees.

Wenner underemphasized the extent to which Frisch, many years earlier, had described extensive experiments which showed that odors are of great importance in recruiting bees to new food sources. He set up a sort of straw man by implying that Frisch claimed that bees *always* dance or that they locate food *only* by information conveyed by the dances. Because bees often locate food by odor, Wells and Wenner denied that the dances convey information about distance and direction under any circumstances. Yet their unwarranted skepticism had the constructive effect of stimulating several new and improved experiments (Esch and Bastian, 1970; Gould et al., 1970; and Lindauer, 1971b). The crux of the issue clearly lies in the behavior of bees stimulated by the dances. Wenner and his associates did point out weaknesses in the experiments by which Frisch had tested the degree to which recruits actually fly to the location indicated by the dances. Many of the results of these experiments could be explained as searching for odors conveyed during the dance.

J. L. Gould (1974, 1975a, 1975b, 1976) has confirmed as conclusively as seems at all feasible that information about direction and distance is indeed conveyed by the dances. He devised a procedure that allows him to alter the direction of the dance so that it describes a different location from that of the actual food source. To do this, Gould took advantage of two details of honeybee behavior that had been discovered by others previously. The first is that if a bright, concentrated light is provided near an observation hive so situated that the bees do not have a direct view of the sky, the bees interpret this light as though it were the sun and orient the dances relative to it, rather than to gravity. Thus, if the food were located 90° to the right of the sun, the dances would ordinarily be pointed 90° to the right of straight up. If the observation hive is inside a small building without windows, and if a bright light is placed to the left of the hive, the bees often orient their dances 90° to the right of this light or approximately upward. Such an artificial light does not seem to alter the efficiency of the dance communication, because both dancers and potential recruits are affected in the same way. The dances are therefore interpreted correctly, even though the reference point is shifted from gravity to the artificial light.

The second detail of bee behavior utilized by Gould is the use of the ocelli—small eyes near the top of the head between the large compound eyes—to monitor the general level of illumination. If the ocelli are covered with opaque paint, the bees behave as though the light level were much lower than it actually is, even though their compound eyes are intact. In the special situation where a bright light inside a shed containing an observation hive causes a reorientation of the dances, covering the ocelli with black paint can prevent this reorientation to the artificial light.

In his experiments, Gould painted the ocelli of foragers with black paint at the feeding station, and an appropriate artificial light was provided near the observation hive where they danced. By carefully adjusting the position and brightness of the light, he was able to avoid reorienting the dances, but shifted the reference point for untreated recruits. The net effect was that the dance communication system could be distorted experimentally so that the dancers pointed toward a location different from that of the actual food source from which they had just returned. The outcome of these experiments was that a great majority of recruits flew to the one of several test feeders at the distance and direction indicated by the experimentally altered dance, and not to the place from which the dancer had actually returned. One should not underestimate the ingenuity with which sufficiently determined skeptics can find some tortuous loophole to provide for an indirect effect of odors (for example, see Rosin, 1980a, 1980b). But the weight of evidence is now overwhelmingly in favor of Frisch's original interpretation. For independent reviews of these questions, see Wilson (1971), Michener (1974), and Hölldobler (1977).

A second sort of skepticism about the "dance language" of honeybees takes a form quite different from that of Wenner. For example, Langer (1972) and W. J. Smith (1968, 1977) accept the conclusion that bees obtain information about distance and direction from the waggle dance, but they interpret the dance as a function of the inner state of the dancer, rather than communication about external objects. On this basis, Langer rejects any conclusion that information transfer by dancing bees is at all comparable to human communication. This and similar opinions will be further analyzed in later chapters.

The views of Terwilliger (1968) exemplify how the tendency to think of animals as Cartesian machines is often based on questionable implicit assumptions. In arguing that language is a uniquely human trait, he dismisses the dances of honeybees as something inferior because of their assumed rigidity: "no bee was even seen dancing about yesterday's honey, not to mention tomorrow's. . . . Moreover, bees never make mistakes in their dance." Leaving aside the technical error that the dances concern nectar, rather than honey. Terwilliger was either unaware of or ignored much of the evidence reviewed above, especially the fact that bees may be stimulated to dance during the middle of the night about a food source they have visited the day before and will almost certainly visit again the next morning. These matters are discussed in more detail in Chapter 8 by Frisch (1967, pp. 349, 352) and Gould (1976).

Premack (1978, p. 628) writes "it is misleading even to label the bee's system language. Appropriate tests of the bee's symbolic capacity appear never to have been made. Suppose a bee gathered information about the direction and distance of food from its hive. As we know, the bee encodes this information in its dance, and other bees can decode the dance. The ability to exchange information in this manner gives the appearance of human language, but the appearance is misleading, as the following test would almost certainly show. When the bee returns, rather than allowing it to dance, show it a dance, and ask it to judge whether or not the dance accurately represents the direction and distance of food. The question is whether the bee can recognize the dance as a representation of its own knowledge. If a bee could judge the relation between the real situation and a representation of that situation, it would be possible to interrogate the bee. . . . If the bee lacks this capacity, then to say it has language is comparable to saying that I have taught this child language except for one small problem: I cannot use the language to interrogate the child. . . ." Premack has added to the definition of symbols as used by Morris and others the requirement that they can be used in a question-and-answer dialogue. In an extention of this argument, Premack (1980) also claims that there is no evidence that bees have "representational capacity."

Certainly the same bees that dance at one time also pay attention to the dances of their sisters a few minutes later. This is perhaps best exemplified by Lindauer's experiments, in which dancers which had visited a good, but not outstandingly desirable, cavity first danced about it, but then became followers of the more enthusiastic dances of others which had visited a better cavity. The same bee can thus alternate between transmitting information by dancing and receiving information by following other dancers. No one has attempted to study exactly what Premack suggests, but it would not be surprising to find that bees react differently to dances that describe the place from which they have just returned, than to dances that convey a different message. For instance, they might recognize dances "synonymous" with their own and avoid them when they change from dancing to following other dances.

To the best of our knowledge, the waggle dances of honeybees are unique in nonhuman animals in the degree to which they exhibit the properties of symbolism and displacement.

They seem so out of place in what we think of as the real world of animals that ethologists tend to ignore them. For example, Krebs (1977) has dismissed them as "an evolutionary freak." But the fact that even one species is capable of so symbolic a type of communication shows that our species does not have an absolute monopoly on such communicative versatility. It is also conceivable that the absence of evidence for anything comparable in other nonhuman animals may result from the difficulty of discovering such flexible communication, given the methods and the patterns of thinking that so far have been available to ethologists. In later chapters I will draw on the extensive data about honeybee communication that has resulted from the work of Frisch and his successors. This is not because I believe that honeybees are uniquely thoughtful, but because better experimental evidence is available about their behavior than we have for all but a few other species. Although nothing comparable to the complexity of the honeybee waggle dances has been reported, another large and abundantly successful group of social insects also engage in communicative behavior which has a greater level of versatility than anyone suspected a few years ago.

Gestural Communication Among Ants

Many species of ants live in complex, interdependent societies that require specialization of function for reproduction, care of developing young, collection of food, and other cooperative behaviors. Some species maintain fungus gardens; others capture and utilize "slaves" from other species. As with honeybees, all the workers that carry out the communication behavior discussed below develop from eggs laid by the single queen in each colony, and are thus sisters or at least half-sisters. In some species, these colonies may number hundreds of thousands or even millions of individuals. Only a few of the hundreds of species of ants have such elaborate organization, but these ant societies have just as great a need for communication to coordinate group activities as do honeybees (Wilson, 1971).

Chemical signals are of primary importance in ant communication. Specialized glands, which occupy relatively large volumes within the small bodies of the ants, produce substances that elicit specific behavioral reactions from other members of the colony. But recent investigations of highly social ants have demonstrated that they also employ communicative gestures. These have been well reviewed by Hölldobler (1977, 1978), one of the leading investigators of social behavior in insects. These highly social species should not be viewed as representative of the entire group of ants; but, like the honeybees, they show what relatively small central nervous systems can accomplish when it has proved important adaptively to communicate with nestmates in order to coordinate complex social relationships.

A number of species of ants lay chemical trails in a specialized type of trail-marking behavior as they return to the nest from places where something important to the colony has been found, usually a source of food. Other worker ants follow these trails from the nest to the food, thus allowing the colony to gather much more food than could be brought in by the individual ants that first located it. These odor trails usually consist of secretion from the hindgut or, in some species, the poison gland. Ants seem unable to determine any directional polarity in such trails, and if they encounter a trail, they are as likely to follow it toward the nest as in the opposite direction. But this limitation does not interfere with the efficiency of the odor trail as a means of guiding additional workers to a food source, because in most cases trail-following is stimulated by some sort of recruitment behavior in or near the nest. Ordinarily, one worker that has located food stimulates her sisters to follow the odor trail she has just laid down. The most specific type of communication occurs in this recruitment of additional workers to leave the nest, follow an odor trail, and perform some appropriate type of behavior at the location where the recruiter found food or something else of importance to the colony.

Until recently, it had seemed that only a single type of general arousal or recruitment was possible, and that ants transmitted no information concerning the nature of the goal to be reached by following the odor trail. As with honeybees, odors brought back by the recruiting ant convey some information about the goal; but because of the difficulties of working with odorous substances which can be effective at very low concentrations, the full details of such chemical communication are not yet understood.

Typical examples of recruitment behavior are provided by such ants as *Novomessor cockerelli*, and *N. albisetosus* (Hölldobler et al., 1978). When foraging workers find a large item of food, such as a dead insect, they release secretions from their poison gland into the air. Nestmates already nearby are attracted from distances up to two meters downwind. In many cases, however, such local recruitment fails to attract enough ants to cut up a large food item and carry the pieces back to the nest. One or more ants then lay a chemical trail with poison-gland secretions as they return to the nest. The odor of this pheromone suffices to stimulate many other workers to follow the odor trail and join in the process of subdividing the food and carrying it back bit by bit.

In other species, the odor of the trail-marking pheromone is supplemented by specific patterns of communicative behavior which increase the likelihood that nestmates will follow the chemical trail and thus be recruited to the food source. For example, Möglich and Hölldobler (1975) describe how *Formica fusca* perform a waggling display after laying an odor trail with material from the hindgut. In this display, one ant faces a nestmate, to which she transfers a sample of the food and vigorously moves her body from side to side. *Camponotus socius* from Florida and *C. sericeus* from Ceylon use a somewhat more complicated type of recruiting behavior (Hölldobler, 1971, 1974). In *C. sericeus* the waggle display, while similar to that of *Formica fusca*, stimulates the recruited ant to grasp and hold on to the rear part of the recruiter's body. The recruiter then runs back along the odor trail with the recruited ant following in what is called tandem running. Similar behavior can be elicited experimentally by attaching the abdomen of a dead ant to a small probe. After suitable stimulation by pheromones, artificially recruited ants can be induced to follow the abdomen and probe as these are moved by the experimenter.

Both species of *Camponotus* move from one cavity to another under many conditions. The old cavity may be too small for an increased population; the feeding territory or even the nest itself may be invaded by more numerous aggressive members of another colony; or scouts may simply have found another cavity with superior physical properties. Under these circumstances, a similar odor trail is laid down, but different recruiting gestures are used. In *C. socius* the motor display used in recruiting to a new nest site involves more jerking, in which the recruiter pulls the head of the potential recruit toward and away from her, rather than from side to side. Thus the recruiting gestures may convey some information about what will be found at the end of the odor trail.

It is not clear why these ants use two different types of recruiting gestures. Perhaps workers vary in their efficiency at these two important tasks and, since ants do not always respond, two messages might recruit more of the required colony members than one generalized signal, which contained no information about what would be found at the end of the odor trail. Or, as the recruited ants travel outward they might prepare themselves in some way for the behavior that will be appropriate when they reach the end of the odor trail. Perhaps they can perform more effectively if they know what to expect.

The African weaver ants *Oecophylla longinoda* present one of the more extreme known examples of complex social organization. These ants construct nest cavities by folding over the edges of leaves and stitching them together. The stitching is accomplished by a worker ant holding one of the larvae, which secretes a sticky material. The larva acts like a shuttle, and by means of hundreds of threadlike connections an effective tent is fashioned. A colony consists of many such enclosures that may be spread over several trees. One colony on Zanzibar included more than half a million individual ants occupying 151 nests spread over 12 trees. But, like other ant colonies, there was only a single queen. Obviously, much coordinated activity is necessary to distribute eggs or larvae from the nest occupied by the queen to all the others, to bring in food, repel invading insects of the same and other species, and to maintain the multiple cavities in appropriate physical condition.

In laboratory experiments, Hölldobler and Wilson (1978) have demonstrated that weaver ants use at least five different recruitment systems to attract sister nestmates from the leaf nests to different goals. Recruitment to food sources includes odor trails produced by the rectal gland together with tactile gestures during face-to-face encounters (opening the mouth, feeling each other with the antennae, and lateral waggling movements of the head). Recruitment of nestmates to new terrain involves similar odor trails, but only tactile stimulation by the antennae. Recruitment to emigrate from the leaf nest to a new site uses quite different behavior patterns, but it is not clear how these are elicited. Recruitment to fight nearby intruders requires a release of a different pheromone as the abdomen is dragged for short distances over the ground. Finally, there is yet another type of recruitment to combat intruders some distance from the nest, and this is mediated by both an odor trail and by specific gestures. These include not only palpatation by the antennae but an intense form of body jerking. The gestures are combined with an alarm communication system of several different pheromones, as described by Bradshaw et al. (1975).

The recruitment of nestmates to join in fighting intruders is particularly significant, both because the gestures used at the initial stages of the recruitment are quite different from those used in recruitment to food, and because they include at least a simple form of secondary communication, in which one ant which has received a message from a returning recruiter continues the recruitment process by repeating the recruiting gesture and thus enlisting additional nestmates.

When *Oecophylla* workers at some distance from the nest

encounter intruders that are too numerous for individual battle, some of them stop fighting, return to the nest, and recruit aid by jerking their bodies rapidly in the direction of a nestmate (several back-and-forth jerks per second). As Hölldobler and Wilson describe it, "This behavior closely resembles the maneuvering of *Oecophylla* workers during territorial fighting . . . to the extent that it can be interpreted as a ritualized version of that more overtly aggressive response." Some of the ants toward which these gestures are directed begin to lay odor trails and "they also exhibited the jerking motor display when they encountered still other workers, even though they had not yet experienced the enemy stimulus themselves. This is a rare example of chain communication in a social insect." In all these cases it should be borne in mind that the gestures are not used in isolation; pheromones and other chemical signals are always involved to some extent, and they, too, may well be more complex than we yet appreciate.

In short, certain species of ants employ communication systems with these two important properties: (1) different gestures are used to recruit nestmates to go out to different goals, specifically food sources, new nest sites, and intruders to be combatted; and (2) in at least one case the ant which receives information from long-range recruitment gestures repeats the recruiting gestures and enlists other nestmates without herself being stimulated by the actual intruders. This type of behavioral chain reaction allows one initial recruiter to attract far more of her sisters to fight against intruders than would otherwise be possible. Although the demonstrated specificity of the communicative gestures is limited to a sort of pantomine, the repetition of the message by ants which have not directly experienced the attacks of intruders is an important feature not previously recognized in animal communication. These two properties suggest at least a rudimentary sort of thinking about the content of the messages.

4

Animal Semantics

Ethologists have observed that several kinds of animals employ different alarm calls when confronted with different sorts of predators. For example, ground squirrels have two types of alarm calls, one for aerial and the other for ground predators (Turner, 1973; Owings and Virginia, 1978; Leger and Owings, 1978; Green and Marler, 1979; Owings and Leger, 1980). Jolly (1972) describes how lemurs and squirrel monkeys also give different calls for aerial and ground predators. But the most convincing evidence available to date that alarm calls may designate specific categories of predators or dangers comes from the carefully controlled field experiments of Seyfarth, Cheney, and Marler (1980). Struhsaker (1967) had observed that East African vervet monkeys *(Cercopithecus aethiops)* give three acoustically distinct types of alarm calls when they see (1) a large mammalian predator such as a leopard *(Panthera pardus)*, (2) a martial eagle *(Polemaetus bellicosus)*, or (3) a dangerous snake, such as a python. The responses of other monkeys differ in an adaptive fashion. The leopard alarm call causes monkeys on the ground to climb trees, the eagle alarm call results in their looking up and running into dense bushes, and the response to snake alarm calls is to look down at the ground, often while standing bipedally on the hind legs, in which posture they can, of course, see a wider area. Seyfarth et al. conducted playback experiments under natural, but carefully controlled, conditions. The responses were analyzed from motion pictures of responses to alarm calls played back from a concealed loudspeaker. Calls of a particular group member in response to an actual predator were played back only when that individual was absent. The results confirmed that these three types of calls elicited the appropriate responses when no predator was actually present, and when all three calls had approximately the same physical intensity.

These three alarm calls of the vervet monkey are noniconic in the sense that they bear no resemblance to the predators they designate or to any sound accompanying the monkeys' responses. They are acoustically distinct in temporal pattern and frequency. Young vervet monkeys give similar calls to a much wider range of external stimuli than do adults, indicating that the exact referents are learned. For instance, infants sometimes give eagle alarm calls to harmless small birds, or even to a butterfly or falling leaf, whereas adults ignore quite similar-looking large birds, such as vultures, which do not attack monkeys.

The distinctive and sensible responses to these three classes of alarm calls seem to show that they convey three distinct messages, namely the presence of one of three kinds of danger. The calls could also be interpreted as injunctions to behave in certain ways. That is, the leopard alarm call might mean "Climb into a tree," or the eagle alarm call "First look up and then dive into the bushes." But in either case these alarm calls convey distinct and adaptively advantageous messages, and thus constitute a simple, but nonetheless significant, example of semantic communication. The calls are arbitrary, in that other sounds could equally well serve the same purpose, and a considerable amount of learning is required to achieve the specificity with which adults employ them to warn their companions of distinct categories of predators. These experiments indicate that vervet monkey alarm calls share one important property with human language, namely reference to external objects and events, a feature that had been judged to be completely lacking in animal communication. Of course, alarm calls convey the information that the caller is afraid; they also inform listening conspecifics what it fears and/or what escape behavior is called for.

Possibly Semantic Communication by Trained Birds

Another possible example of communication about specific external referents may be provided by the report that a great spotted woodpecker *(Dendrocopos major)* learned a simple telegraphic drumming code by which it requested whichever of five types of food it wished to obtain from the experimenter at a particular time (Chauvin-Muckensturm, 1974; Chauvin and Chauvin-Muckensturm, 1980). When the woodpecker had learned to use this drumming code to communicate with the experimenter, other persons with whom the bird was not familiar were able to communicate with approximately equal effectiveness.

These controls do not entirely eliminate the possibility of a "Clever Hans error," named for a horse, Clever Hans, which had apparently learned to carry out complex arithmetical communications and transmit the results by tapping with one foot. Although many scientists were convinced of the genuineness of this accomplishment, more careful study showed that the horse was actually watching the person who presented the problem by writing numbers on a blackboard, and who had to count in order to determine whether the horse was giving the correct answer (Pfungst, 1911). In the course of such counting, the person performed small movements unconsciously, and it was these which Clever Hans had learned to notice. He had also learned to stop tapping when the experimenter stopped indicating his own process of counting by nodding or otherwise, or when he signalled inadvertently that he expected something to happen. Because new human observers also gestured in minor ways during their counting, without realizing that they were doing so, Clever Hans was able to perform his apparent feats of arithmetic even for strangers. Sebeok (Sebeok and Umiker-Sebeok, 1980; Umiker-Sebeok and Sebeok, 1980) claims that this sort of error confounds most or all experiments in which animals are judged to communicate symbolically. Despite these limitations, the experiments of Chauvin-Muckensturm are, at the very least, suggestive, and deserve to be extended and elaborated to determine whether they can be replicated.

An African gray parrot named Alex has recently been trained to use English words to request objects that he seemed to enjoy playing with (Pepperberg, in 1981). His functional vocabulary includes nine nouns: paper, key, wood, "hide" (rawhide chips), "peg wood" (wooden clothes pins), cork, corn, nut, and "pasta" (bow tie-shaped pieces of macaroni). He also uses three color adjectives: rose (red), green, and blue, and two words to describe shapes: three-corner (triangle) and four-corner (square). When presented with any of 30 familiar objects, Alex produces the correct two-word combinations (color-noun or shape-noun) with an accuracy of greater than 80 percent. Accuracy is reported to be at least as great on the first as on later presentations. He has also learned to use a recognizable "no" in place of his normal raucous squawk when rejecting an undesired object or when protesting against a procedure he dislikes.

Previous attempts to train parrots to use correctly their recognizable imitations of human words seem to have failed because food was used as the reward for correct behavior (Mowrer, 1950, 1952, 1954, 1958). Pepperberg's experiments appear to have succeeded because she used objects in which Alex was already interested, things he liked to bite and play with. This work is one of several recent experiments demonstrating that some stimuli can be far more easily associated than others with particular responses. The experiments are still in a preliminary stage, but the controls for Clever Hans errors seem reasonably adequate. They illustrate how easy it is to conclude that versatile animals like parrots are not capable of simple thinking, on the basis of what appeared in the heyday of behaviorism to be wholly adequate tests that led to negative results.

Signing Apes

Several well-known studies of gestural communication between chimpanzees and human experimenters have had a wide and profound impact, for they suggest that these close but non-human relatives are capable of symbolic communication much more like human language than anything previously thought to be possible. Several earlier attempts to teach chimpanzees to make vocal sounds were significant in their almost total failure. Even after years of effort, home-reared chimpanzees learned to produce only a very few approximations of monosyllabic words, although they recognized many words of human speech (Hayes, 1951; Hayes and Hayes, 1951). The Gardners (1969, 1971, 1975*), stimulated in part by Goodall's (1968) observations of wild chimpanzees, decided that gestures were a more promising method for communication, as Robert Yerkes had suspected might be the case (Bourne, 1977). They trained a wild-born young female chimpanzee, Washoe, to use several dozen "words" from the American Sign Language for the deaf. An important part of their procedure was the total immersion of Washoe in a social environment consisting of human companions who communicated only in this sign language while in her presence. In four years, Washoe acquired approximately 130 signs, invented a few of her own, and used them all in conversational exchanges with her human companions. In "blind" experiments, she was able to name pictures presented by an experimenter who could not see them himself. Washoe spontaneously used signs and sometimes "signed to herself" when alone. She transferred at least a few of the signs appropriately to new situations. For example, the sign for "open," which she originally learned for doors, she later used to request the opening of boxes, drawers, briefcases, and picture books.

Washoe learned to use gestural signals much as words are used by very young children, but of course many differences remain between her signing and early human speech. For example, word order seems to play a much smaller role, if any, in Washoe's signing than it does with children who have vocabularies of comparable size. There is no convincing evidence that signing apes have developed rule-guided patterns in which signs are combined to give new meanings to the combinations. Investigations of gestural communication by chimpanzees have been continued both by the Gardners (1979), Wood et al. (1980), and by Fouts, Lemmon, and their colleagues at the University of Oklahoma (reviewed by Fouts and Rigby, 1977). Among many significant findings, their studies have demonstrated that, to at least a limited extent, Great Apes can communicate with each other by means of a sign language they have been taught by human experimenters. Patterson (1978) has also trained a gorilla to use signs in a manner quite similar to Washoe's signing.

Chimpanzees can also learn to identify objects and pictures on hearing their names in spoken English. This ability allowed Fouts et al. (1976) to train a three-year-old male chimpanzee named Ally to utilize both spoken English and sign language. Ally acquired a vocabulary of more than 70 reliable signs and also learned to understand several spoken phrases and words. He was then taught new signs corresponding to 10 of the spoken words to which he was already responding correctly. These were names of familiar objects, but the objects were not present during this phase of the training. After training was completed, Ally showed himself completely capable of using the gestural signs to identify correctly the objects for which they stood.

Premack and others have studied the languagelike behavior of chimpanzees by different types of experiments (Premack, 1976; Premack and Woodruff, 1978; Rumbaugh, 1977; Savage-Rumbaugh et al., 1978, 1980). These utilize a relatively small number of symbolic objects or mechanical devices which the chimpanzees learned to use appropriately. In experiments with a chimpanzee named Sarah, Premack used colored plastic tokens as names for familiar objects, and Sarah learned to use these tokens to request specific items of food. She also learned to use the tokens correctly even when they bore a completely noniconic and arbitrary relationship to the objects they represented.

In experiments by Rumbaugh and his colleagues at the Yerkes Laboratory, the chimpanzee uses a keyboard to request desired objects or simple actions. In these experiments, the vocabulary is limited to symbols or keys provided by the experimenter and, although relatively large repertoires have been built up and used correctly, the experimental situation greatly limits the possibilities for the animal to acquire a large vocabulary and to generate new "words" spontaneously. Although these experiments can be more rigorously controlled, they offer the chimpanzee less scope for originality than do the methods used by the Gardners, Fouts, or Patterson. These differences in experimental approach are less important than the fact that both approaches have yielded similar results: chimpanzees have learned to use surprisingly large vocabularies of gestures or manually manipulated symbols to communicate far more complex messages than scientists had previously believed were possible in any nonhuman animal.

The details and significance of languagelike communication learned by chimpanzees have been extensively discussed and reviewed, for example by Klima and Bellugi (1973), Linden (1974), Thorpe (1972b, 1974a, 1974b), Bronowski and Bellugi (1970*), S. J. Gould (1975), Ristau and Robbins (1979, 1981a, 1981b), Seidenberg and Pettitto (1979), Terrace (1979), and Terrace et al. (1979). One interpretation is that the basic ability to communicate is severely limited by the anatomy and physiology of the chimpanzee larynx. This seems less likely than the alternative hypotheses that the chimpanzee brain is capable of relatively complex communication, but that this capability can be expressed far more readily through manual gestures than by vocalization. Extensive observations by Goodall (1968, 1971, 1975) have clearly demonstrated that wild chimpanzees communicate with considerable effectiveness by means of gestures and facial expressions. Yet these have proved to be so difficult for human observers to analyze in detail that we cannot say whether they convey anything more than such emotional states as threat, affection, hunger, or sexual enthusiasm.

Studies by Savage-Rumbaugh et al. (1977) reveal that gestural communication can be used spontaneously by pygmy chimpanzees to convey information about motions a male wishes a female to make, or positions he wishes her to assume, in preparation for copulation. In other experiments at the Yerkes Laboratory, two young male chimpanzees, Sherman and Austin, have been trained to request from one another simple tools they have learned to use for obtaining foods (Savage-Rumbaugh et al., 1978, 1980).

During the past two or three years, several critical discussions and improved experiments have thrown additional light on the languagelike behavior that can be taught to captive Great Apes. Terrace et al. (1979) raised a young male chimpanzee named Nim Chimpsky for approximately four years, during which they taught the animal, by methods similar to those of the Gardners, to use a considerable number of signs patterned after American Sign Language. Unlike earlier experiments, extensive video tapes were made not only of Nim's signing, but of the behavior—and especially the signing—of the human trainers with whom he interacted. Unfortunately, due to a lack of adequate facilities and resources (always a most serious problem in this type of work), Nim's training had to be divided among a large number of people, only a few of whom worked with him long enough to establish as close a social rapport as that attained by the Gardners and their colleagues with Washoe and, more recently, with other young chimpanzees.

Terrace and his colleagues agree with the Gardners that chimpanzees can learn vocabularies of a hundred or more signs and use them more or less like single words to obtain or refer to specific objects or actions which are of interest to them. But Terrace et al. are very skeptical of earlier claims that this type of behavior has much in common with what they regard as the essential features of human language. In reaching this conclusion, they place great emphasis on grammar and the rule-governed patterns by which we combine words into sentences. Nothing of this sort was observed in Nim's signing.

Terrace et al. (1979) summarized their work in a paper entitled "Can an Ape Create a Sentence?" Perhaps a better title would have been "Did Nim Learn to Use Sentences?" My point is that it is far from clear that the extensive and laborious training procedures used with Nim or with other chimpanzees would necessarily be expected to lead to the use of grammatical sentences. Regardless of this question, however, it remains very clear that the signs which captive Great Apes have learned to use have not yet been shown to represent more than the equivalent of single words. But, as discussed in more detail in Chapter 5, Descartes, Chomsky, and many others believe that "the word is the sole sign and certain mark of the presence of thought."

Serious questions have also been raised by Terrace and his colleagues, especially by Seidenberg and Petitto, by Ristau and Robbins, and even by Savage-Rumbaugh and Rumbaugh, as to whether the languagelike behavior taught to apes is truly comparable to human language. One alternative interpretation is that the apes have learned to use certain signals in order to obtain particular objects or elicit particular patterns of behavior from their companions. This is regarded as problem-solving rather than language. As discussed in Chapter 8, the essence of the distinction is that whereas human language is very often used to solve problems and elicit desired behavior from companions, it is also used for other, more cognitive, purposes. These include, but certainly are not limited to, attempts to change the beliefs of listeners rather than simply trying to obtain a given behavior from them. The distinction has been well put by Bennett (1978). In discussing experiments in which apes have been trained to use one or another form of languagelike behavior, Bennett suggests that the performances of these animals "are analogous to *injunctions*—requests, commands, pleas, etc.—which aim to elicit *behaviour* from the other party. [We should] contrast injunctions with *statements*, which aim to produce *belief* (or awareness, or realization, or knowledge) in the other party. . . . I think that in most human communication the speaker does intend to produce in the hearer a belief-change which may, but need not, be intended to have some specific further behavioural upshot."

In view of our very limited knowledge of whatever awareness or thinking may occur in nonhuman animals, it is difficult to ascertain the degree to which apes or other animals intend to produce a change in the behavior of conspecifics or other animals, rather than merely seeking to acquire some desired object, such as food. Yet sometimes animals clearly request particular kinds of behavior from others, for example, grooming. Premack and Woodruff have begun to develop methods which offer some hope of detecting and analyzing the beliefs of chimpanzees, but it is too soon to expect significant results, and the limited data so far presented are open to many different interpretations (Savage-Rumbaugh et al., 1978, 1980).

Like Terrace et al., linguists tend almost universally to dismiss the languagelike behavior of captive apes as wholly different from human language. This is primarily because they feel it to be not only very much simpler and more limited in vocabulary, but to lack both the spontaneity and creativity of human language and anything approaching grammar. It is difficult to judge the degree of spontaneity which has actually been exhibited by the apes, because in many experiments their opportunities for spontaneity are severely limited, as when they have available only a fixed set of plastic symbols or computer keys.

When gestures similar to sign language are used, there is of course a considerable opportunity for an ape with a vocabulary of, say, one-hundred such "words" to use them in new ways or in new combinations. The published reports of such accomplishments are very few in number, given the time that has been available to the experimenters. This suggests that new utterances are far less frequent than one would expect if lively and inquisitive animals such as chimpanzees were really capable of using their newly acquired communication system to express the variety of thoughts and desires of which one would judge them capable from the versatility, and even from the mischievousness, of their overt behavior.

When some complex behavior, such as signing by chimpanzees, is acquired only after long and arduous training, and especially when close social companionship and rapport with human trainers or experimenters is necessary, the question of artificiality arises. Closely linked to this question is the danger, discussed in vigorous detail by Sebeok and Umiker-Sebeok (1980), that Clever Hans errors, or other types of inadvertent cueing by the human experimenters, produces behavior which is not understood by the animal but is erroneously interpreted by the deeply involved human trainers as meaningful and intentional communication. These dangers are less acute when we are dealing with natural communication systems used by animals in their normal social behavior. No one has seriously suggested that Frisch or others have inadvertently cued honeybees to perform their waggle dances.

This raises the question of the relationship between the artificial, languagelike behavior taught captive apes and their natural communication systems. Seidenberg and Petitto (1979) cite as a reason for rejecting any similarity to human language the claim that many of the signs learned by captive chimpanzees are actually quite similar to communicative gestures used by wild chimpanzees. This does not seem to be a serious objection, for if the Gardners and their successors have simply elaborated on a natural system of gestural communication, it would mean that such communication is, to that extent, not a laboratory artifact, but part of the real world of chimpanzees under natural conditions.

Menzel (1974, 1978), Menzel and Halperin (1975), Menzel and Johnson (1976), have shown that captive chimpanzees can communicate fairly complex information by some combination of gestures or expressive movements that human investigators have not yet deciphered. In one of their experiments, several chimpanzees who were familiar with each other were confined temporarily in small cages at one edge of a large outdoor enclosure they used habitually. One animal then was led to something such as food, not visible from any of the isolation cages, shown the object, and returned to his cage. Next, the entire group was released, and the "leader" was able to convey the location of the hidden object rapidly and efficiently. In these experiments, a cognitive map or some sort of internal representation of the outside world seems to be communicated from one animal to another. Sometimes, when the leader seemed not to wish his companions to discover the object, he appeared to attempt to keep them from locating it. These apes seem capable of conveying or withholding information from their companions intentionally. If these experiments had been conducted with silent human beings, an observer would have had no doubt that the leader knew where the object was located and either did or did not wish his companions to find it.

We are caught here in the sort of dilemma that often arises when one tries to analyze complex and interactive behavior. Carefully controlled experiments are necessary to ascertain with confidence what information is actually conveyed by any sort of communication, but laboratory conditions may be so unnatural as to distort or obscure whatever sorts of thinking the animals would do under natural conditions. If the system is at all complex, observing it under natural conditions may not suffice to interpret it fully, any more than it would suffice to watch a television program in which the actors came from a wholly different human culture and used a language of which we were totally ignorant. It is for reasons of this kind that it is natural to turn to experimental interchanges with animals suspected of communicating, and perhaps thinking, in more complex terms than we are accustomed to consider. These possibilities will be discussed in more detail in Chapter 10.

5

Is Man Language?

The linguist Bloomfield (1933) recognized that animals do communicate, but concluded: "Human speech differs from the signal-like actions of animals, even of those which use the voice, by its great differentiation. Dogs, for instance, make only two or three kinds of noise—say, barking, growling, and whining. . . . When we tell someone, for instance, the address of a house we have never seen, we are doing something which no animal can do." The molecular biologist Monod (1975) reiterated the widespread view that "man is endowed with a competely unique capacity, which no other species shares, namely language. . . . There is nothing argumentative for instance, in animal communication." This opinion overlooks the many cases where animals exchange ritualized threat signals and can reasonably be considered to be arguing about who should retreat. And before swarming honeybees have reached a consensus about the cavity to which they will emigrate, as discussed in Chapter 3, their conflicting dance messages have many attributes of an argument. Presumably Monod was using the term argumentative to mean exchanges of more complex and reasoned statements.

Many philosophers and linguists have also argued that human language is closely linked with thinking, if not identical and inseparable from it (Cassirer, 1953*; Fodor et al., 1974; Hattiangadi, 1973; Healy, 1971; Lenneberg, 1971; Pyles, 1971; Thass-Thienemann, 1968; Weiss, 1975). Langer has expounded this view in several contexts (1942, 1967, 1972) and with special eloquence and vigor (1962): "language is symbolic, when no animal utterance shows any tendency that way. The biological factors that caused this great shift in the vocal function were, I believe, the development of visual imagery in the humanoid brain, and the part it came to play in a highly exciting, elating experience, the festal dance." This was written only a few years before Goodall (1968, 1971) described what appeared to be highly excited "rain dances" of chimpanzees, in which the adult males of a social group respond to the first heavy rains by violent displays that include loud pant-hoots, rushing about, and breaking off branches from trees. Langer continued: "As I remarked before, images are more prone than anything else we know to become symbols. . . . In animals typically, every stimulation that takes effect at all is spent in some overt act. . . " (Langer, 1962). "A genuine symbol is, above all, an instrument of conception, and cannot be said to exist short of meeting that requirement; that means that an ape thinking symbolically could think of an act he had no intention or occasion to perform, and envisage things entirely remote from his real situation. . . . Symbolism is the mark of humanity" (Langer, 1972). These views may have been expressed before the full impact of the Gardners' breakthrough concerning chimpanzee communication had been felt, and before the experiments of Menzel and Halperin (1975). But it is now clear that some animals communicate complex messages so closely attuned to the nuances of the social situation that great caution is called for in reaching such definite conclusions as those expressed by Langer.

Black (1968) assured us that "It would be astounding to discover insects or fish, birds or monkeys, able to *talk to one another* . . . [because] . . . Man is the only animal that can talk . . . that can use *symbols* . . . the only animal that can truly *understand* and *misunderstand*. On this essential skill depends everything that we call civilization. Without it, imagination, thought—even self-knowledge—are impossible." The neurologist Critchley (1960) was so impressed by human speech that he wondered: "Can it be, therefore, that a veritable Rubicon does exist between animals and man after all? . . . Can it be that Darwin was in error when he regarded the differences between man and animals as differences merely in degree?" Goldstein

(1957) asserted in the same vein that "Language is an expression of man's very nature and his basic capacity. . . . Animals cannot have language because they lack this capacity. If they had it, they would . . . no longer be animals. They would be human beings." To Anshen (1957), "Man *is* language."

The influential contemporary philosopher Noam Chomsky, in his penetrating discussions of the nature of language (Chomsky, 1966, 1972), subscribes to the tradition widely attributed to Descartes. To both Descartes and Chomsky, language is the essence of humanity. In a blend of translation and eloquent reiteration, Chomsky (1966) ably summarizes the Cartesian view that no men are "so depraved and stupid, without even excepting idiots, that they cannot arrange different words together, forming of them a statement by which they make known their thoughts; while, on the other hand, there is no other animal, however perfect and fortunately circumstanced it may be, which can do the same . . . man has a species-specific capacity, a unique type of intellectual organization which cannot be attributed to peripheral organs or related to general intelligence and which manifests itself in what we may refer to as the 'creative aspect' of ordinary language use—its property of being both unbounded in scope and stimulus-free. . . . Human reason, in fact, is a universal instrument which can serve for all contingencies, whereas the organs of an animal or machine have need of some special adaptation for any particular action . . . no brute [is] so perfect that it has made use of a sign to inform other animals of something which had no relation to their passions . . . for the word is the sole sign and the only certain mark of the presence of thought hidden and wrapped up in the body; now all men . . . make use of signs, whereas the brutes never do anything of the kind; which may be taken for the true distinction between man and brute."

Chomsky continues, "The unboundedness of human speech, as an expression of limitless thought, is an entirely different matter [from animal communication], because of the freedom from stimulus control and the appropriateness to new situations.

Modern studies of animal communication so far offer no coun-
terevidence to the Cartesian assumption that human language
is based on an entirely different principle. Each known animal
communication system either consists of a fixed number of signals, each associated with a specific range of eliciting conditions
or internal states, or a fixed number of 'linguistic dimensions',
each associated with a non-linguistic dimension." The evidence
reviewed in Chapter 3 calls into serious question these sweep-
ing, negative generalizations of Descartes and Chomsky.

It is difficult to ascertain whether the communication behavior of any particular animal consists of an absolutely fixed number of signals or to establish just what eliciting conditions or
internal states are associated with each. Thus Chomsky's assertion of numerical fixity must remain simply an unsupported
opinion. It seems quite possible that the perceived rigidity and
limitation to a few specific conditions or states exists in the minds
of human commentators rather than in the real world of animal
behavior. Human language and the thoughts that it expresses
are not infinite in their scope and inventiveness, as implied by
Chomsky and others. If they were, we might not need such
elaborate educational systems to develop our mental and linguistic abilities. Animal behavior, and in particular animal communication, is also adaptable to new situations and is even creative, at least under some circumstances. Consider, for example,
how Mike, a Gombe Stream chimpanzee, used kerosene cans to
enhance the acoustical component of the charging displays by
which he displaced the previously dominant males of his social
group (Goodall, 1971, pp. 112–117). Mike's use of a wholly new
type of noise-making object in intermale encounters showed
every sign of being an intentional effort to improve his social
status, and seems to have been a creative, rather than a stereotyped, behavior pattern.

Price (1938) argued that if animals use symbols, we must
assume they have minds. Bee dances are certainly symbolic, but
Chomsky (1972) maintains that one cannot trace similarities and
evolutionary continuities between animal and human commu-

nication. "When we ask what human language is, we find no striking similarity to animal communication systems . . . human language, it appears, is based on entirely different principles. This, I think, is an important point, often overlooked by those who approach human language as a natural, biological phenomenon; in particular, it seems pointless, for these reasons, to speculate about the evolution of human language from simpler systems. . . . As far as we know, possession of human language is associated with a specific type of mental organization, not simply a higher degree of intelligence. There seems to be no substance to the view that human language is simply a more complex instance of something to be found elsewhere in the animal world."

Cultural transmission of human language has often been cited as one criterion establishing it as unique to our species. For example, Pollio (1974*) states three criteria necessary to qualify an event as a symbol: it must be representative of some other event, "freely created," and transmitted by culture. The dances of honeybees are recognized as being representative, but are held to be too rigid and unvarying to satisfy the second criterion, and to be genetically programed rather than culturally transmitted.

The influential views of Chomsky also include a belief that the capability for learning and using language is a species-specific human attribute. Although he does not say so explicitly, it is difficult to escape the conclusion that Chomsky feels there must be a substantial genetic component in the development of our ability to use language; this is strongly implied in the way in which he refers to it as species-specific. Pollio's and Chomsky's position on this species-specificity in *Homo sapiens* is a bit ambiguous, because they also seem to believe that culture is another unique human attribute. Culture has been defined in many ways, but basically it is used to mean a shared set of learned behavior patterns for regulating social interactions. Some, but not all, anthropologists' definitions imply that people who share a culture consciously understand the rules they have learned in order to regulate social behavior.

Lorenz (1969) has reviewed the considerable evidence that cultural transmission is important in the social behavior and communication of birds. Sarles (1975) has reviewed the difficulty of basing a rigid human-animal dichotomy on the criterion of language. It has recently become apparent to several thoughtful ethologists and others that many social animals learn the kinds of behavior which enable their societies to function effectively. There seems to be a real, though relatively simple, type of culture present in many nonhuman animals, especially primates and songbirds, as described in some detail by Bonner (1980) and Mundinger (1980). We have tended to take it for granted that only people are consciously aware of the social relations that make up their culture, and that nonhuman animals, no matter how much they may learn to cooperate, are unthinking automata. But accumulating evidence makes it almost impossible to defend rigorously any qualitative dichotomy.

This brings us back to the question of rigidity and total genetic programing in the behavior of social insects, honeybees in particular. Under some conditions, the waggle dances do vary considerably, and they are not performed at all unless certain social conditions are present. As I have mentioned earlier, there is a great deal of communication among members of a hive of bees, largely through exchange of stomach contents and transmission of chemical signals. This serves to regulate the activities of the workers and stimulate some of them to search for particular materials when these are in short supply. Part of this social regulation consists of the reception a returning forager receives when she tries to exchange food with one of her sisters. If the material regurgitated is not accepted by the other bee, the returning forager has difficulty finding a taker for her stomach contents and is less likely to seek more of the material. This process of social regulation has been well studied in hives that become overheated; this causes a marked change in the behavior of the foragers. Instead of searching for and bringing back nectar or other concentrated sugar solutions, they search for water, which cools the hive as it evaporates (Frisch, 1967; Lindauer, 1971a).

Furthermore, the specific dance patterns used to convey information about the location and quality of a given source of food must be learned and remembered by the individual worker on each occasion. The counterargument is that relatively rigid genetic instructions cause worker bees to behave in ways which serve this communicative function, even though individual elements of the behavior are influenced by learning. Although it is clear that bee-dancing is under much stronger genetic control than is human language, the two communication systems have important qualitative elements in common. The prevailing view of insect behavior holds it to be controlled entirely by genetic programing, or at least that insects are programed to learn only certain things under particular conditions. Bees are thus viewed as complex automata equipped with "on-board computers" that have been genetically programed by natural selection to cope with all eventualities (Gould, 1979). The postulated programs must of course provide for rare but important situations, like the need to seek out and report the location of new cavities at the time of swarming. But can people learn absolutely anything, and do we learn equally well under all circumstances? Our patterns of thinking about other species place such great emphasis on genetic control of their behavior that we tend to slip very easily into unqualified assertions.

The view that insects are genetically programed automata is so widely and deeply accepted, even by ethologists, that it is extraordinarily difficult to accept the implications of the versatile communication system discovered in honeybees by Frisch. It almost seems as though one must choose between the "robot" view of insects and what would otherwise be overwhelmingly convincing experimental evidence that honeybees communicate with each other in a flexible manner capable of dealing with any of several different subjects, including new problems of great importance. The very fact that we would readily accept the bee dances as evidence of intentional communication if they had been discovered in primates, rather than in insects, should warn us that our frame of reference may not be sufficiently flexible to take advantage of truly revolutionary new discoveries.

It is also worthwhile to consider the overwhelming mutual interdependence of such social insects as honeybees. No individual can survive for any extended period in isolation, and reproduction is entirely dependent on an elaborate series of behavior patterns in which nonreproductive animals contribute directly and indirectly in numerous ways to the feeding, protection, cleaning, and other aspects of the behavior of the queen or reproductive females. Furthermore, the development of each individual egg is very different from the preprogramed growth and differentiation of an isolated zygote. At many stages during larval and pupal stages, older bees feed and clean the developing egg, larva, or pupa, and protect it through numerous activities necessary for maintaining the colony. Even the sex of an individual bee is strongly influenced by chemical materials supplied by other members of the genetically related colony. To the best of our knowledge, the communicative dances of a given genetic strain of honeybee always take a very similar, if not precisely identical, form, but species-specificity does not necessarily mean total control by the genotype. Social influences during the lifetime of the individual could well have some effect, as Schneirla (1966), Lehrman (1953), Hinde (1970), Gottlieb (1971), and others have argued for birds and mammals.

In short, we may be skating on thin ice when we assume that everything about the behavior of social insects results directly from a chain of causal sequences beginning with their DNA and proceeding immutably toward rigidly stereotyped adult behavior in total isolation from any influence of the physical or social environment. Such arguments would be more plausible if the egg developed in a wholly isolated situation and was not so abundantly and directly influenced by care-taking behavior. In very general terms, it seems clear that genetic instructions affect the ability of the adult organism to learn a particular type of communicative behavior. As in almost all nature-nurture considerations, there are good reasons to believe that both genetic and environmental influences are of major importance in shaping the adult animal and its behavior. The same considerations can be applied to honeybee dances and, as emphasized by Chomsky (1966, 1972), to human language.

Of course, no one in his senses can overlook the enormous differences in complexity, subtlety, and versatility that separates human language from any known, or even speculatively postulated, communication between members of other species. But most scholars and scientists concerned with the question have not been content with quantitative distinctions—differences in degree rather than differences in kind. Thus, Hockett (1958) made an effort to formulate objective criteria by which human language can be qualitatively distinguished from animal communication.

Hockett's original list has been modified by Hockett and Altmann (1968) and by Thorpe (1972a, 1974a, 1974b) in an attempt to encompass the variety that exists in animal communication and yet to distinguish features unique to human language. Although the task seems to become increasingly difficult as more and more is learned about communication in other species, it is important to review the 16 design features included by Thorpe (1974a) in the latest version of this general scheme: (1) use of the *vocal-auditory channel;* (2) *broadcast transmission and directional reception;* (3) *rapid fading;* (4) *interchangeability* (the same individual can act either as transmitter or receiver of information); (5) *complete feedback* (the organism emitting the signal also perceives everything relevant about the message); (6) *specialization* (relatively weak signals trigger biologically important consequences); (7) *semanticity* (the communication system is used to correlate and organize the life of a community); (8) *arbitrariness* (signals or symbols are abstract, in that the meaning they convey is independent of their physical properties); (9) *discreteness* (signals are unitary entities and do not grade continuously into one another); (10) *displacement* (discussed in Chapter 3); (11) *openness* or *productivity* (meaning that new messages can readily be created and understood); (12) *tradition* (conventions passed on from one generation to the other by learning); (13) *duality of patterning* (while single units of the communication may be meaningless, patterned combinations of them convey important information); (14) *prevarication* (using communication signals to convey information known to be inaccurate); (15) *reflectiveness* (the ability to communicate about the communication system itself); and (16) *learnability* (the ability of a user of one communication system to learn another one employed by a different group of organisms). All of these features are certainly present in human language, and the question that arises is the degree to which any of them, or any combination, provide an objective basis for concluding that there is a fundamental difference in kind between human language and all communication systems used by other animals.

Most of the 16 design features are, in fact, present in many animal-communication systems. These include *reliance on the vocal-auditory channel; broadcast transmission and directional reception; rapid fading; interchangeability* (animals can act both as transmitters and receivers); *specialization* (energy in the signal small compared to the effects triggered by it); and *complete feedback* (transmitting animal able to perceive all relevant properties of his signal). Another set of design features seems, at first thought, to be distinctively human, but similarities are certainly present in many animals. These include *semanticity*, defined as use of signals to correlate and organize the activities of a community on the basis of associations between the signals and properties of the surrounding world. Many animal communication signals certainly satisfy this criterion in a general way. For example, territorial songs of birds and the social communication of primates correlate in an important fashion with the properties of the environment as far as conspecifics are concerned. Chemical and gestural communication is crucial in coordinating the mutually beneficial activities of social insects.

Arbitrariness is another criterion that falls into this category. Bee dances often are considered not to be arbitrary because the dance pattern is a sort of iconic replica of the route to be flown. But there are so many other aspects of the dances, such as their vigor and the role of sounds of vibrations in conveying something akin to the urgency of the message, that it becomes little more than a matter of semantics whether to designate these features as arbitrary. For example, the use of "up" as a point of reference meaning toward the sun is arbitrary; the system would work equally well if "down," or "90 degrees to the right of up," meant toward the sun. To be sure, all honeybee colonies use this same convention, in contrast to human languages, where a given meaning is often conveyed by quite different words.

Eight other design features are more difficult to find outside of human language. *Discreteness* is an important property of human linguistic communication, in that small elements, such as words or syllables, do not functionally grade into one another. But the definition of discreteness depends heavily upon the size of element considered. For example, a single cycle of the honeybee dance or even a single cycle of abdomen waggling could well be considered a discrete unit. The latter, in particular, is combined in various ways with other elements, such as sound pulses. Jostling and bumping seem to elicit a rather generalized state of arousal in many insects. But in the waggle dance, as discussed in Chapter 7, individual movements have been combined into an organized pattern which conveys a much more specific meaning: "this way, and this far."

Two other features—*tradition*, the meaning of signals transmitted by teaching and learning, and *learnability*, users of the communication system learning about it from another—are closely related and can best be considered together. It is clear that learning and social tradition play a large role in the details of bird song and other types of social communication (Smith, 1977; Green and Marler, 1979; Mundinger, 1980; and Bonner, 1980). Bee dances are generally considered to be genetically programed, but here, too, the details are certainly learned, as when bees visit and dance about a location conveyed to them by

other dancers. It is difficult to disentangle genetic and environmental effects in the development of complex behavior in animals that live in such interdependent societies as those of honeybees, as critically discussed by Marler et al. (1980).

Another design feature frequently stated to be lacking in
animal communication systems is *duality*. A system is said to
have duality if signal elements are meaningless in themselves
but become meaningful when formed into appropriate combinations. Here, again, the applicability of the criterion depends
upon the size of unit considered. Bee dances or other forms of
communication behavior can easily be subdivided into individual elements, such as single muscle contractions, which by
themselves have no communicative significance. Human language obviously achieves a great deal of its enormous importance by use of compound and complex combinations of small
elements; but we do not know enough about animal communication to judge the degree to which combinations, as opposed to
individual signals, may be important.

One design feature often considered unique to human language is *displacement*. As discussed in Chapter 3, displacement
means that the communication process can refer to things remote in time or space. Clearly, bee dances and the recruiting
gestures of weaver ants satisfy this criterion. Another similar
criterion is *openness*, meaning the ease and frequency with
which new messages are coined by using previously unused
combinations of elements of the communication system. This is
sometimes also called *productivity*. Ever since Frisch's first decoding of the waggle dances, it has been obvious that they often
concern locations and kinds of food about which the bees have
never danced before. And when swarming bees exchange reports about the location and desirability of the cavities they have
located, the whole subject matter of the communication is a
brand new one for the individuals involved. Beer (1975) believes
that some calls used by gulls are "semantically and pragmatically
open," as will be discussed in Chapter 10.

The fifteenth criterion, *reflectiveness*, the ability to communicate about the communication system itself, is a relatively recent addition to the list. Thorpe feels this property "is undoubtedly peculiar to human speech," yet we should ask ourselves whether, if it does occur in animals, any of our available methods of investigation would suffice to disclose it. The discussion of self-awareness in Chapter 2 is pertinent to this issue. Although it was not mentioned in this context by Hockett and Altmann, *prevarication* is one more criterion commonly advanced to set our species apart from other animals. I will discuss it further in the next chapter.

Thorpe accepts the available ethological evidence, especially the studies of chimpanzees by the Gardners and by Premack, as convincing evidence that apes, at least, and probably also dogs and wolves, clearly demonstrate conscious purposiveness. He feels it is likely that, if the chimpanzee larynx were adequate, these apes could learn to speak as well as children three years old, or perhaps older. To Thorpe, "human speech is unique only in the way it combines and extends attributes which, in themselves, are not peculiar to man but are found also in more than one group of animals. . . . Yet . . . there comes a point where 'more' creates a 'difference'. Here he aptly quotes A. N. Whitehead (1938): "The distinction between men and animals is in one sense only a difference in degree. But the extent of the degree makes all the difference. The Rubicon has been crossed."

It is only fair to point out that many of the opinions discussed above date from the "pre-Washoe" period of ethology, and might not reflect the considered views which these authors would now hold. Yet there is no doubt that for centuries philosophers and linguists have based their most fundamental definitions of humanity on very positive assertions about what animals can and cannot do. This means that whatever students of animal communication have learned, or can learn in the future, about communication behavior is directly relevant to major questions of fundamental significance to linguistics and philosophy.

6

Are Animals Aware
of What They Are Doing?

Insofar as linguists and philosophers have been correct in linking human thinking so closely to language, as reviewed in the previous chapter, the communication behavior of other species is bound to suggest conscious thought roughly to the extent that it shares essential features with human speech. In allowing ourselves to entertain the notion that animals may be aware of past, present, and future events, or may experience mental images, in the sense discussed in Chapter 2, it certainly is not necessary to assume that such mental experiences are at all similar to those which a person might have under analogous circumstances. Yet almost all linguists, and most philosophers who have considered the question, have vacillated between denying to animals any significant mental experiences (for example, Langer, 1942, 1962*, 1967, 1972*), and grudgingly admitting the likelihood of certain simple ones while rejecting others, which are then held to be of crucial importance.

These discussions are often eloquent; but they show signs of what ethologists call conflict behavior. To some, this will appear to reflect a fundamental difference between scientists and humanists, but I am more optimistic, and suggest that communication behavior presents a magnificent opportunity for fruitful interaction and cross-fertilization between broad-minded scientists and equally perceptive humanists (Griffin, 1978a). The implicit denial of mental experiences to animals has become almost an act of faith, and it is supported primarily by arguments and assertions that true language is a unique and characteristic attribute of our species.

It seems that our own thinking and behavior as scientists is sometimes influenced by a feeling of "own-group superiority." We find it easy to believe, and to seek evidence supporting the belief, that our own group, and in particular our own species, is enormously superior to all others. This conviction is prominent in western Judeo-Christian social philosophies, and we use it to justify exploiting other species for our own benefit. Perhaps widespread and deep-seated feelings and beliefs of this kind have resulted to a considerable extent from natural selection. If animals are capable of any beliefs at all, natural selection may operate on these beliefs as on other phenotypic characters, as discussed in Chapter 9. It seems plausible that animals would be more likely to survive and reproduce if these beliefs included confident faith in their own superiority and the assurance that exploiting other species was normal and correct behavior. Of course, if one believes that only our species is capable of any sort of belief or awareness, this argument is irrelevant; or it may seem applicable to human evolution, but not to that of any other species.

All of these viewpoints present difficulties because pertinent evidence is almost nonexistent. Nevertheless, it may be helpful to allow ourselves the luxury of speculating about the adaptiveness of a belief in the superiority of one's own group. Such a belief could well reinforce behavior patterns that lead to energetic efforts to dominate, exploit, prey upon, displace, or otherwise take advantage of other species. Could it be that even as highly intellectual, language-dependent scholars and scientists we are still swayed to some extent by such a deep-seated sociobiological influence? The psycholinguist Roger Brown (1958) opens a discussion of the comparative psychology of linguistic reference with a light-hearted but pointed paraphrase of current opinion: "I grant a mind to every human being, to each a full stock of feelings, thoughts, motives, and meanings. I hope they grant as much to me. How much of this mentality that we allow one another ought we to allow the monkey, the sparrow, the goldfish, the ant? Hadn't we better reserve something for ourselves alone, perhaps consciousness or self-consciousness, possibly linguistic reference?

"Most people are determined to hold the line against animals. Grant them the ability to make linguistic reference and they will be putting in a claim for minds and souls. The whole phyletic scale will come trooping into Heaven demanding immortality for every tadpole and hippopotamus. Better be firm now and make it clear that man alone can use language and make reference. There is a qualitative difference of mentality separating us from the animals."

Later (pages 164–171), Brown recognizes that "If vocalization is acknowledged to be unimportant, the dances of the bee appear to be very much like referential language." But he places great emphasis on the assumption that "the dances are unlike language in that they are not learned." He feels that animal communication is rigid, always predictable when the circumstances are specified, whereas human speech is not. He says of the dancing bees "the followers' reaction is *too* reliable."

The impression of mechanical predictability of the responses of follower bees stems in part from the eloquent simplicity of Frisch's descriptions of this behavior in his semipopular books and articles. More detailed reading of his technical papers, or actual observation of the bees themselves, show that there is quite enough variability in the behavioral responses of bees and other communicating animals to leave room for the assumption of spontaneity. Many followers do things other than flying out to the place indicated by the dances; they often seem to ignore the dances altogether and turn to other activities. Even those that do leave the hive do not all reach the indicated goal. It is technically so difficult to observe individual bees known to have followed a given dance, once they are flying in the open air, that we know almost nothing about their behavior between the time they leave the hive and the moment they arrive at the feeding place. As pointed out by J. L. Gould (1975a, 1976), bees newly recruited by dances often take much longer to reach the food than the time necessary for a direct flight.

Examples of "philosophical conflict behavior" are evident when Kenny, Longuet-Higgins, Lucas, and Waddington (1972, 1973) discuss the problems of analyzing mental processes from their respective viewpoints as philosopher, physicist, theologian, and biologist. When they consider the possibility that nonhuman minds might exist, they reflect the current climate of opinion by devoting much more attention to computers than to animals. At one point, Washoe and other chimpanzees are denied true minds on the ground that they merely mimic the sign language of the deaf, but elsewhere because they have been taught this language by human trainers. But, as discussed in Chapter 4, I suggest that the important question is not how signing apes acquire their communicative behavior, but how they use it, and what they think they are doing. Kenny et al. seem to agree that animals have subjective feelings and perceptions, that is, sense-consciousness. These scholars express a viewpoint similar to that of Natsoulas (1978a, 1978b) who makes more explicit the distinction between perceptual awareness, which he calls "consciousness$_3$," and awareness of oneself, which he designates "consciousness$_4$."

Waddington argued that "if consciousness were to be adopted as a criterion of mind, it would be a signally useless one, because the only way to tell whether any other thing is conscious is to ask it. And that you can only do to human beings; . . . the concept of consciousness is not applicable to anything but a language using animal." Yet Waddington recognized that some animals, such as dogs and cats, are capable of having intentions, and Longuet-Higgins admitted that: "An organism which can have intentions I think is one which could be said to possess a mind [provided it has] . . . the ability to form a plan, and make a decision—to adopt the plan." From this divergence of opinions, the presence of mental images and an ability to provide introspective reports on self-awareness and intentions emerge as criteria of mind.

Hampshire (1959) clearly expressed the opinion of many

philosophers: "It would be senseless to attribute to an animal a memory that distinguished the order of events in the past, and it would be senseless to attribute to it an expectation of an order of events in the future. It does not have the concepts of order, or any concepts at all. An intention involves, among other things, a definite and expressible expectation of an order of events in the future, and is possible only in a being who is capable of at least the rudiments of conceptual thought." How can Hampshire or anyone else be certain that no nonhuman animal has concepts of any kind or concepts about the order of events? Leaving aside for the moment the question of expectations about future events, memories of past patterns and events often involve temporal relationships. Animals learn to perform certain actions not in randomized combinations but in a definite sequence. Indeed, learned motor behavior would otherwise be chaotic and ineffective.

Hampshire is clearly mistaken in denying that any nonhuman animal can remember the order of events and that none is capable of any sort of concept. An enormous body of evidence gathered primarily by comparative psychologists shows that monkeys, rats, and pigeons can learn to respond correctly to relationships between stimuli. The animal is first trained to solve a particular kind of problem and then tested with new stimuli that share certain relatively simple relationships with previously learned examples. Out of numerous examples of this general sort of learning reviewed by Mackintosh (1974) and by Hulse et al. (1978), a few are especially pertinent. Apes and monkeys can learn to match newly presented items to a sample presented previously on many occasions. They also can learn, after rather extensive training, to select from three objects presented to them the one which differs from the other two (Harlow, 1949).

Another type of complex learning has been called "transposition" by psychologists. This involves the learning of relationships, rather than responses to specific objects. For example, Köhler (1918) reported that both chimpanzees and chickens could be trained to select the lighter of two gray cards and then make the correct relational choice when presented with a new pair that differed in darkness. In some experiments the darker of the second pair might be identical to the lighter of the first pair, thus indicating that it was the relationship, rather than the absolute value of the stimulus, that had been learned. Although behavioristic psychologists struggled for about 20 years to avoid the conclusion that an animal could understand even as simple a concept as relative brightness, experiments such as those of Lawrence and DeRivera (1954) and others reviewed by Mackintosh (1974) and Hulse et al. (1978) indicate that animals really do learn the relationship rather than the properties of individual stimuli. A recent example of the serial organization of behavior, which was of such concern to Lashley (1951), has been reported by Straub et al. (1979), who have demonstrated that pigeons could learn to select a sequential pattern of four colors. The colors were presented as a row of disk-shaped windows illuminated from behind. The pigeons learned to peck these disks according to their color (yellow, red, green, blue, for instance), regardless of their relative positions, which were varied randomly.

We must therefore suppose that Hampshire and others who hold similar views really mean that, although animals are perfectly capable of learning to perform definite sequences of behaviors and also can respond correctly to particular sequences of stimuli they have experienced in the past, they are somehow unable to think conceptually about these relationships. But how can we distinguish between appropriate behavior or response, on the one hand, and conceptualization of the relationships to which the animal responds or which govern its behavior? Perhaps the emphasis should be placed on Hampshire's requirement that expectations of future events be *expressible*. Of course, such expectations are even more difficult to detect than are organized memories, unless one accepts as evidence intention movements or injunctions such as the bee dance, which certainly seem to refer in some way to future behavior. In this regard, one could even accept Smith's (1977) interpretation of the bee dance as a complex intention movement that conveys information only about the future behavior of the dancer and still see evidence that it involves, in Hampshire's words, "an expressible expectation of an order of events in the future."

Pigeons can separate highly varied pictures of natural scenes and objects into two categories, after learning that responding to one category produces food. The criteria are based on the pictured presence or absence of water or of a particular person (Herrnstein et al., 1976). In similar experiments, leaves of one species of tree were distinguished from those of other trees (Cerella, 1979). This ability suggests that pigeons entertain simple concepts such as "food is available where water can be seen, or where a particular woman is visible, or where the leaves are those of white oaks, but not of other trees." It is also possible that we are dealing simply with a surprisingly versatile type of categorical perception. It would be interesting to learn whether pigeons or other animals could classify motion pictures of certain types of action when performed by a variety of animals or people under a wide range of conditions.

Other sorts of evidence strongly indicate that chimpanzees and other animals can understand at least moderately complex concepts. One example is provided by the laboratory experiments of Rohles and Devine (1966, 1967) on the "middleness" concept. A chimpanzee was trained to select from a row of identical objects the one in the middle position, in the sense that equal numbers were to its left and right. If the objects were arranged symmetrically, the chimpanzee could, after prolonged training, select the middle one when as many as 17 were present. If irregularly and asymmetrically arranged—that is, with irregular spacings between adjacent objects—it could solve the problem when up to 11 were presented. This is roughly equivalent to the ability of four- to six-year-old children to solve comparable problems. Apparently rhesus monkeys and pigs can learn to solve this type of problem only with seats of three or five. Perhaps Rohles' and Devine's chimpanzee learned to solve these problems without acquiring a concept that corresponds to our understanding of middleness, for instance by learning each odd-numbered arrangement independently of all the others. A critical test might require that, after learning to select the middle member of an odd-numbered set, say seven, the animal could, without further training, select the middle member of another set that contained a different odd number of objects. For example, if trained to select the middle one of three, seven, and eleven, would the animal also pick the middle member out of five or nine without previous training with sets of five or nine? Unfortunately, this type of experiment apparently has not been carried out.

The extensive field studies of ethologists have also cast grave doubt on negative assertions of Hampshire and many others. For example, ethologists have found the term "intention movement" widely applicable to those postures and relatively slight movements of animals that convey to other animals reliable information about probable future behavior (Daanje, 1951; Tinbergen, 1951). There is no doubt from the reactions of conspecifics that such information is indeed conveyed. But it has been the curious custom of most ethologists to stop short of interpreting an intention movement as evidence that the animal has a *conscious* intention, although Daanje did state that animals intend to perform particular behavior patterns. Since both conspecifics and human observers can predict the future behavior of an animal from its intention movements, it seems remarkably unparsimonious to assume that the animal executing the intention movement cannot anticipate the next steps in its own behavior.

These reductionist attitudes are highly resistant to erosion, however, and even the broadly perceptive biological commentator Lewis Thomas (1974) writes: "A solitary ant, afield, cannot be considered to have much of anything on his mind; indeed, with only a few neurons strung together by fibers, he can't be imagined to have a mind at all, much less a thought. He is more like a ganglion on legs." Ignoring the fact that worker ants are females, may we ask whether a dancing honeybee is a ganglion on wings?

Bennett (1964) argued with charming erudition that to qualify as rational creatures bees would have to exchange more abstract messages than real bees had been demonstrated to do. But Bennett's ideas about real bees were apparently based on Frisch's first reports of his discoveries. Bennett did not consider the implications of the later finding that the dances are used to communicate about things other than food or to exchange information and reach a consensus about hive sites. As discussed by J. L. Gould (1976), recent observations have shown that bee dances do satisfy many of Bennett's criteria for rationality, whereas others have never been looked for by observations or experiments that could reveal their presence or absence.

Black (1968), McMullan (1969*), Robinson (1973*), and others have distinguished animal communication systems from human language on the ground that the former are rigid responses to external or internal stimuli, which can, at least in principle, be definitely specified, whereas human language is spontaneous, creative, and unpredictable. As pointed out above in connection with the views of Brown (1958), animal responses to communication signals are, in fact, highly variable. We know so little about animal thinking that it would be very difficult to detect new and spontaneously created signals if they did occur from time to time, as discussed in Chapter 8.

Maritain (1957) exemplifies this climate of opinion: "Animals possess a variety of means of communication but no genuine language . . . no animal knows the relation of signification or uses signs as involving and manifesting an awareness of this relation . . . in the last analysis . . . the relation of signification remains unknown to the bees. They use signs—and they do not know that there are signs. . . . The whole thing belongs to the realm of conditioned reflexes, whereas language pertains to the realm of the intellect, with its concepts and universal notions."

Grice (1957) stated a widely held view of the psychological conditions necessary for genuine language: "Perhaps we may sum up what is necessary for A to mean something by *x* as follows. A must intend to induce by *x* a belief in an audience, and he must also intend his utterance to be recognized as so intended." As discussed in Chapter 4, most ethologists are still unsure whether nonhuman animals communicate with conscious intent to alter the behavior of social companions. Their beliefs, if any, are almost inaccessible to our investigation, so that we cannot yet take up the question whether they have beliefs about the beliefs of others.

It is commonly stated that no animal can use its communication system to tell a lie. Of course, a lie requires attention to deceive, so that to judge whether variability in animal communication behavior is "noise" or prevarication requires knowledge of the animal's intentions. But conscious intention is a category of mental experience that is so widely believed to be uniquely human that ethologists seldom consider it.

Monkeys and apes have been observed to make efforts that seem designed to secure food without attracting the notice of more dominant members of the social group, who would otherwise be very likely to secure the food themselves. Rüppell (1969) described cases in which a mother Arctic fox was competing for food with her several well-grown young; the latter resorted to such drastic competitive tactics as urinating in their mother's face in order to reach food morsels first. After several such encounters, the mother often gave warning calls, otherwise used to signal dangers of various kinds and, when the youngsters ran off, seized the food herself. It is difficult to interpret such behavior without postulating at least short-term intentions and plans on the parts of both mother and young. The behavior patterns they used to compete for especially tasty food would ordinarily have been applied to quite different situations.

A serious tendency toward circularity exists in all these arguments. Conscious intention in animals is ruled out *a priori* and then its absence is taken as evidence that animal communication is fundamentally different from human language. It seems that the more difficult the question under consideration and the less adequate the available evidence, the more definite become the generally accepted assertions about the differences between human language and animal communication, as diagramed in Figure 1. But it is important to ask on what basis such definite assertions can be made about what bees and other animals do or do not know. Have we allowed nonscientific value-judgments to color our thinking about these questions?

For example, Adler (1967) concludes that the communication system of honeybees is "a purely instinctive performance on their part and does not represent, *even in the slightest degree*, the same kind of highly variable, acquired or learned, and deliberately or intentionally exercised linguistic performance that is to be found in human speech." He then goes on to argue that if it were to be established by some future investigations that animals differ from men only in degree and not radically in kind, we would then no longer have any moral basis for treating them differently from men. Indeed, this matter has already arisen. A gorilla named Koko was borrowed from a zoo as an ailing infant, raised by Patterson (1978), and taught sign language by procedures similar to those used with Washoe and other chimpanzees. It has been reported that when the zoo asked for the return of this valuable animal, a strong counterargument was made. Because she had learned language, it was now morally wrong to treat her as a zoo animal. Apparently the problem was resolved by donations of funds sufficient to reimburse the zoo, but the ethical issue was raised in definite and explicit form.

Adler also makes the converse argument that if it should be established that animals differ from man only in degree and not radically in kind, such knowledge would destroy our moral basis for holding that all men have basic rights and an individual dignity that render it wrong to mistreat groups of men judged to be inferior for the benefit of supposedly superior groups. Followed to its logical conclusion, this argument implies that the comparative investigation of communication behavior has more dangerous potential consequences than nuclear physics had in the 1930s, or the current fear that synthesis of certain new forms of DNA might produce uncontrolled pathogens (Berg et al., 1974; Goodfield, 1977). Full acceptance of the Darwinian revolution may have a considerable psychological impact insofar as the distinction between mental experiences of animals and men comes to be recognized as one of degree rather than one of kind. Unless we wish to abandon our scientific faith that understanding fundamental processes will prove of value to our own species, I suggest taking Adler's philosophical arguments as one more reason why it is important for scientists to investigate, as fully and accurately as we can, the relationship between human and animal communication and cognition.

7

Evolutionary Continuity
of Mental Experience

Despite the deep reflection and thoughtful eloquence of philosophers, the nature, and even the existence, of mental phenomena have remained largely outside the scope of natural science. Occam's razor has cut so deep since the 1920s that behavioral scientists have grown highly uncomfortable at the very thought of mental states or subjective qualities in animals. When they intrude on our scientific discourse, many of us feel sheepish, and when we find ourselves using such words as fear, pain, pleasure, or the like we tend to shield our reductionist egos behind a respectability blanket of quotation marks.

There have been a few exceptions to the behavioristic tradition and the related stress on parsimonious explanations in biology. Adams (1928) discussed what he felt were logical weaknesses in Morgan's original statement of his widely cited canon ("In no case may we interpret an action as the outcome of the exercise of a higher psychical faculty, if it can be interpreted as the outcome of the exercise of one which stands lower in the psychological scale"). Adams found Morgan's arguments "insufficient to criticize the inference of mind." One problem with Morgan's canon is that it is based on an intuitive classification of behavior and "psychical faculties" into higher and lower categories, the latter to be preferred for reasons of parsimony unless the evidence forces postulation of the former. But no definitive, objective reasons have yet been provided for assigning behavior to particular places on the scale of lower to higher. Adams advocated that mental experiences could reasonably be inferred in another animal to the degree that its "structure, situation, history, and behavior" resemble those which accompany such mental experiences in a human observer. Buytendijk and his colleagues have attempted both theoretical and experimental analyses of the subjective experiences of animals. These have been reviewed by Thines (1977).

Jennings (1906, 1910, 1933) repeatedly and eloquently argued the case for an open mind concerning behavioral and mental continuity between men and other animals. Marler (1974) believes that "we delude ourselves if we think that a complete [behavioral] discontinuity separates us from other animals." And Boyle (1971) advocated that "the psychologist . . . will, if he is wise, acknowledge that organisms make sense of their experiences, and he must attempt to discover what this sense is. Unfortunately psychology has turned its back on this task because, since a psychologist's hypotheses about an animal's experience cannot be confirmed, psychologists have to a large extent ceased trying to understand other human beings as well." Yet Boyle, although ready to admit that when a cat rubs against his feet she intends to induce him to give her food, nevertheless found it "difficult to imagine that a bee *intends* to communicate to fellow workers a message about the distance and direction of pollen. . . . That may be the meaning of the dance to an observer, but it is doubtful whether bees are capable of this type of understanding." Small size or phylogenetic remoteness from man are evidently taken as evidence against any form of conscious intent in bees. The image of a "ganglion on legs" dominates our view of invertebrate animals.

Tolman (1932) developed a "purposive behaviorism," in which what are ordinarily considered as mental events and processes were treated as intervening variables between external stimuli or internal influences on the one hand and overt, observable behavior on the other. By simply designating mental processes as intervening variables, the difficult question of their nature is avoided. But Tolman's position was less rigid than strict behaviorism, and he accepted the reality of conscious awareness in animals, for example in a white rat at the moment of learning some new behavior pattern, such as a specific portion of a maze. In recent decades, however, most psychologists and ethologists have either avoided this question or taken positions closer to strict behaviorism.

Many schools of philosophy dissent vigorously from materialism and from logical positivism, and not only do these schools accept the reality of concepts which behaviorists reject as meaningless; they often attach central importance to them. But philosophers of this kind seldom pay much attention to animals. There is an active discipline of cognitive psychology which has felt free to deal with *human* mental experiences, although often refraining from the use of explicit mentalistic terms (Mowrer, 1960a, 1960b; Taylor, 1962*; Fodor, 1975; Estes, 1975; Natsoulas, 1978a, 1978b; Underwood and Stevens, 1979). Neisser (1967) asserted emphatically that "Cognitive processes surely exist, so that it can hardly be unscientific to study them." Schultz (1975) reviewed the trend of some contemporary psychologists to abandon the strict taboos of behaviorism. When Irwin (1971) considered many of these broad and challenging questions within a fairly conservative framework of objective analysis, he did recognize that men and animals have "expectancies." By this term, he seemed to mean something closely resembling mental or internal images of the possible future outcomes of various alternative patterns of behavior. The term consciousness, however, was still one which Irwin struggled to avoid. But Kimble and Perlmuter (1970*) went so far as to speak of volition. Fodor et al. (1974) reviewed the contributions of psycholinguists who have followed the lead of Chomsky in rejecting a behavioristic position regarding *human* language, and found it essential to consider mental entities. But few psychologists or even ethologists have yet moved away from an essentially behavioristic position with regard to animal behavior.

Lorenz (1958, 1963) has been a notable exception among ethologists; he does not hesitate to express a belief that animals have subjective experiences, although, like Adams, he has concentrated his attention on higher vertebrates. Bertrand (1969) hesitantly ventured to speak of a monkey's behavior as voluntary. Razran (1971) briefly considered the possibility that animals have simple thoughts. Brewer (1974) has argued in the course of a vigorous dissent from behaviorism: "Since cognitive theory holds for humans, it is unparsimonious not to apply it to animals." And Weiner (1972*) stated that "cognitivists also use the man-animal continuity to promote their view that even infrahuman behavior is guided by cognitive processes." Mowrer (1960a, 1960b) struggled to escape the rigid restrictions of the behavioristic position while still dealing only with observable events, but was nevertheless led to state that "if consciousness were not itself experienced, we would have to invent some such equivalent construct to take its place." Yet he concludes that bee dances are "limited to 'sentences' of the thing-sign variety" because samples of food are transferred, rather than being represented by a symbol. But, as explained in Chapter 3, the dances can serve as symbolic communication in the absence of any samples of food.

Wittgenstein (1953) approached these problems with Socratic questions, such as: "We say a dog is afraid his master will beat him; but not, he is afraid his master will beat him to-morrow. Why not?

"One can imagine an animal angry, frightened, unhappy, happy, startled. But hopeful? And why not?

"A dog believes his master is at the door. But can he also believe his master will come the day after to-morrow?—And *what* can he not do here? Can only those hope who can talk? Only those who have mastered the use of language?"

The essence of Wittgenstein's skepticism seems to concern the time span of an animal's anticipation into the future, and he may well be correct in this estimate for a dog in the situation suggested. But suppose a dog *did* anticipate events likely to occur tomorrow. How could we recognize this fact from observing its behavior today? Lacking any evidence at all, we make a negative judgment, reasonably enough, but tend to forget its weak foundation. And what basis do we have for denying that any nonhuman animal can hope? For example, how do we know that hungry animals never hope that the next potential food source they visit will, in fact, yield some nourishment?

Miller (1967) asserted that "Man is the only animal to have a combinatorially productive language . . . a species-specific form of behavior. . . . Serious attempts have been made to teach animals to speak. . . . These attempts have uniformly failed in the past and, if the argument here is correct, they will always fail in the future." Rensch (1971*) and Popper (1972*) have expressed similar views. Recent successes in teaching chimpanzees to communicate with gestures clearly cast doubt on the validity of Miller's "pre-Washoe" prediction. As discussed in Chapter 4, it seems clear that a variety of gestures or other artificial forms of communicative behavior can be used by chimpanzees in a fashion analogous to single words. Nevertheless, the most recent studies of languagelike behavior in apes have brought into serious question previous claims that *combinations* of signs have been used in a consistent and meaningful fashion. Thus Miller's emphasis on the property of being "combinatorially productive" remains close to the heart of the human-animal distinction. But however limited may be their combinatorial productivity, the signing apes certainly seem to be expressing at least simple thoughts.

It may be that all species but our own use single signals as entities which convey only one meaning at a time, and that combinations of them never take on an important new meaning. On the other hand, the whole pattern of honeybee communication, of which the waggle dances are a central part, has many attributes of a meaningful combination. The trophallactic exchange of food and accompanying scents constitute one component which is widespread among social insects. Lateral wagging of the abdomen when one insect is excited is another component, and in many other species it serves to arouse and stimulate nestmates to food-finding activity. Dethier (1957) found one of the clearest examples of this in blowflies. When one blowfly has found a rich source of food, it may execute such dancelike gestures. These serve to stimulate its hungry fellows to search about much more vigorously than they would otherwise have done. This sort of communicative behavior is clearly much more global and nonspecific than the waggle dances of honeybees, and there is no evidence that information is conveyed about distance or direction. Finally, the locomotion of a number of insect species is influenced in much the same way both by gravity and by the location of a bright light. When on a level surface they tend to move toward a light. If they are then placed on a sloping surface in darkness or diffuse light, they often move upward. These patterns of orientation are elaborated in honeybees into an integrated pattern that serves as a simple form of symbolic communication.

Yet even the capacity to learn a human language is widely believed to be under strong genetic control (Chomsky, 1972, 1976). Perhaps the crucial criterion of conscious intention is the adaptation of a genetically based capacity to newly arisen needs, as in the case of swarming honeybees. It can be argued that the integration and combination has occurred in evolution, rather than being freely and spontaneously created by an individual animal. But when viewed from a sufficiently broad perspective, the difference between our species and the social insects may not be as great and as nearly absolute as we have been accustomed to believe.

Miller et al. (1960), Langer (1962, 1972), and others state without qualification that man is the only animal that can be aware of his own future death. But I suggest that we pause and ask just how anyone knows this. What sort of evidence is available either pro or con? Suggestive inferences can be based on the clear demonstration that many social animals recognize each other as individuals, and on the observation that some animal mothers show signs of distress over the corpses of their dead infants, which they carry about for days (Goodall, 1968, 1971, 1975). How can we judge whether an animal may experience any notion of its own future death after observing the death of companions (Cowgill, 1972)? The available, negative evidence supports at most an agnostic position.

Evidence from Neurophysiology

Eccles (1973, 1974) has proposed an explanatory framework for brain function that includes conscious experience, which he feels exists only in the human dominant cerebral hemisphere. He cautiously reserves judgment concerning the existence of consciousness in the subordinate hemisphere of the human brain and in the Great Apes. Sperry (1969) went further, and not only recognized the importance of conscious experience in the human brain, but concluded that it "exerts a directive holistic form of control over the flow pattern of cerebral excitation." Much of the evidence on which Eccles and Sperry have based their ideas comes from surgical cutting of the corpus callosum in human patients suffering from severe epilepsy. When the hemispheres are thus deprived of their normal channel for exchanging information, they can learn to recognize quite separate sensory stimuli, and they seem to operate almost independently with respect to complex behavior. This applies both to human patients and to all other mammals tested so far. In some patients, only the dominant hemisphere can report verbally about its sensations and learned behavior. But in others, mental functions seem more evenly divided between the hemispheres (Nebes and Sperry, 1971; Teng and Sperry, 1973; Gazzaniga, 1975, 1979; Levy, 1979).

In some of the patients with a severed corpus callosum, the subordinate hemisphere can mediate learning to recognize objects, and the patient can demonstrate that learning has occurred by pointing to the correct object when asked to match a sample. Yet the same subject is unable to report the correct choice by speaking or writing. It seems clear that the subordinate hemisphere carries out many of the mental functions ordinarily considered conscious, but lacks the ability to report them in words. To Eccles (1974), "the minor hemisphere resembles an animal brain." These discoveries add to the evidence for physiological continuity between men and animals in brain function, and suggest a comparable continuity in mental experiences. Pribram (1971) and Gazzaniga (1975) review much the same evidence and reach conclusions similar to those of Eccles, although they are expressed in slightly different terms.

Does Behavioral Complexity Imply Conscious Awareness?

Many behavioral scientists express feelings of discomfort, or even outrage, at the inference of conscious intention in animals because previously unsuspected complexities in their orientation and communication have been discovered. On strictly logical grounds, complexity of behavior and conscious awareness are neither commensurate with, or necessarily related to, one another in any way (Gould, 1979). Inanimate mechanisms can be enormously complex and difficult to understand, but most *descriptions* of animal behavior can be modeled by mechanisms far simpler than a television receiver. The same can be said of many physiological mechanisms. For instance, a very simple electronic circuit can produce an electrical signal that closely resembles the spike potential of a neuron. But only the most naive engineer-turned-neurophysiologist would accept the existence of such a circuit as a satisfactory explanation of the functioning of nervous systems. To Loeb, the existence of a phototactic machine constructed out of wheels, electric motors, and photocells was evidence for believing that animal, and even human, behavior could be explained in terms of tropisms or forced movements. But the crippling limitations of such intellectual myopia should now be clearly apparent; the simplicity often lies not in the behavior, but in its description.

Despite the above considerations, it remains a fact that one of the principal reasons that led linguists and cognitive psychologists to abandon the strict behavioristic approach to human language and cognition was the staggering complexity of stimulus-response mechanisms that had to be postulated in order to account for human behavior on the basis of Skinnerian formulations (see, for instance, Chomsky, 1959). Of course, it is always possible to postulate more and more complex and intricate stimulus-response relationships, but those to which one is driven in seeking to encompass human thinking and human conversations become so unwieldy that they can scarcely be justified on the grounds of scientific parsimony.

As more and more is learned about the social behavior and other adaptive responses of animals to unpredictable environments, a similar problem arises. Behaviorists, however, can still set up stimulus-response models of almost all *known* examples of animal behavior. It may become almost a matter of taste when one reaches a sort of tolerance threshold for compounded layers of stimulus-response formulations. Our climate of thinking encourages a higher threshold for other species than for other people. It should also be borne in mind that virtually the entire body of information available from ethology and comparative psychology has been gathered by scientists more or less strongly committed to a noncognitive approach and to reductionist or stimulus-response explanations. This could scarcely be expected to bias the available body of evidence in favor of cognitive interpretations.

Can we accept the reality of our own conscious awareness but reject the hypothesis that any pattern-recognizing machine is also consciously aware of the pattern to which it responds selectively? It certainly seems easier to reject the notion of consciousness in a simple mechanism, such as a lock built to accept a particular key, than in a computer system programed to respond correctly to an especially intricate input signal. This is primarily because we can understand how the lock works, but find it impossible to encompass in our own mental imagery at any one time the complexities of the entire computer program. Must we therefore infer conscious awareness whenever we do not know how a mechanism works? Complexity of some kinds surely provides no convincing evidence for the existence of mental experiences. The physiological mechanisms by which a kidney regulates the chemical composition of the blood or the biochemical systems that regulate cellular respiration are marvelously complex and incompletely understood, but certainly are far different from central nervous systems in their structure and function. Analogies can be drawn by describing all three in terms so general that they apply to any self-regulating system, and these system properties are of interest in their own right. But brains and minds, insofar as we allow ourselves to admit the existence of the latter in animals, are surely different in basically significant attributes from kidneys and mitochondria (Sperry, 1973*).

In the interpretation of communication behavior, pride of parsimony can lead us into some awkward situations. As mentioned briefly in Chapter 3, W. J. Smith (1975, 1977) has argued that the waggle dances of honeybees convey to other bees not information about distances and direction, or about the actual location of a distant food source, but only information about the internal state of the dancer. According to Smith, "A forager bee does not state that it has found food at a certain place; rather, she describes a direction of flight (perhaps a flight she is likely to make again shortly) and, on request, provides a sample of the food." But suppose we knew nothing about human language and watched and listened to human beings while they were conversing. We might well conclude that they, too, were simply describing their internal states. Indeed, the definition of internal states can easily be extended to cover the most complex speech or writing, which can, if one wishes, be interpreted as "merely describing the internal state" of speaker or writer. Pushed to their limits, all these arguments make sense only on the implicit assumption of conscious intent on the part of human beings and its absence in animals.

The general feeling that our species is uniquely superior has suffered a series of intellectual setbacks that began with the Copernican and Darwinian revolutions. Later, the ability to learn from past experience was advanced as a unique human attribute, but successive discoveries of learning in animals more and more distantly related to us forced abandonment of that criterion. Tool-using suffered a similar fate as a result of studies of Darwin's finches, for example, and of sea otters, as did tool-making more recently at the hands of chimpanzees (Goodall, 1968; McGrew, 1974; McGrew et al., 1979; Warren, 1976; Spuhler, 1977). These are only a few of many examples reviewed by Beck (1980) in which animals use a suitable object which they select, or sometimes fashion, from their surroundings to satisfy some need which would otherwise be beyond their capabilities. Certain ants thrust bits of leaf or wood into soft or semiliquid food materials and then carry the loaded "sponge" to their home colony, where the tool-user and her sisters consume the food thus gathered. The same piece of wood or leaf may be used repeatedly for this purpose, and the efficiency of transport is increased substantially over what the ant could otherwise bring back. Some satin bowerbirds hold a wad of soft bark in the bill to aid in decorating their bowers with a semiliquid coloring material that is formed in the bird's mouth from fruit pulp, mud, or other material mixed with saliva (Marshall, 1954). Vultures throw stones at ostrich eggs with sufficiently good aim to break an egg about half the time and thus obtain a rich source of food. Certain finches in the Galapagos Islands detach twigs or cactus spines, hold them in the bill, and probe into crevices to pry or force out insects they could not otherwise dislodge. These probes may be improved before use by breaking off excess length, and they are carried in flight from where they were obtained to the crevices where they are used.

Sea otters carry stones under their armpits and use them as anvils against which to pound shellfish they have brought to the surface. They do this while floating on their backs, holding the stone in one hand and the hard-shelled prey in the other. They have also been observed to carry similar stones to the bottom and use them to pound abalone shells until they can be torn away from the rocks.

Beck reflects the contemporary *Zeitgeist* by coupling his

review of these instances of tool use with the argument that such behavior is no stronger evidence of intelligence than are other types of problem-solving or cooperative social behavior. But tool use is not only a striking and important category of flexibly adaptive behavior; it is one which seems to require that the tool-user, the tool-carrier, and especially the tool-maker, must make intentional plans to achieve certain goals by a set of motor actions not combined into similar coordinated patterns for any other purpose.

When evidence is presented that a nonhuman species achieves some of the criteria that previously had been proposed for distinguishing human language, "the list grows longer in order to exclude the interloper species. If this kind of progression continues we may eventually have a definition of language that isomorphically maps the behavior of human language; in essence a redundant description of the behavior" (Fouts, 1973). Similar viewpoints have been presented in semipopular fashion by Linden (1974) and S. J. Gould (1975).

For many years, most biologists concerned with animal behavior—for example, Dobzhansky (1967) and Simpson (1964)—seemed to share the basic views of the linguists and philosophers quoted above. The reductionist, behaviorist tradition dominated the thinking of ethologists until very recently. When Frisch or Lindauer discovered that bees dance in patterns correlated with the location of food or something else needed by the colony, or that a bee which had danced about one potential location for the swarm shifted to a different site under the influence of the more enthusiastic dances of others, they did not suggest that perhaps the bees feel any need for sugar, pollen, water, or a suitable new cavity for the colony. That was taboo (Lindauer, 1955*, pp. 312–313). Virtually all descriptions and discussions were in terms that would be equally applicable to a living animal or an appropriately contrived machine.

This viewpoint served our science well for more than 50 years by constraining speculations and focusing attention on phenomena amenable to experimental analysis. Seventy years ago, this disciplined restraint was a healthy reaction to an earlier tendency to ascribe human feelings to a wide variety of animals, mostly on the basis of rather unconvincing anecdotal evidence. But suppose the biologists and psychologists active at the turn of the century had known about the communicative dances of bees and about the recent insights the Gardners, Premack, the Rumbaughs, Terrace, and others gained from Washoe and her successors. Would they have been so adamant in banishing from respectable consideration all notions of mental states in animals? Their message was: As a working strategy of research, assume no mental states or subjective experiences, and see how much of animal behavior can be accounted for on this parsimonious basis. This has now been done on a large scale, and some of the results show that it is time to review our perspectives and strategies in the light of the new discoveries.

Intentional Communication

The importance of the questions discussed in this book is demonstrated by the heavy reliance of linguists and philosophers on the consciously intentional use of language as the principal distinguishing characteristic of our species. A major reason for this philosophical assertion has been the acceptance by those linguists and philosophers of the general conclusions expressed by students of animal behavior. I suggest that behavioral scientists now have the opportunity, and perhaps an obligation, to explore and discuss the limitations of this traditional, behavioristic viewpoint in the light of recent discoveries about communication behavior in animals.

When the behavioristic position is stated at its scholarly best—for example, by Lashley (1923, 1958)—it is essentially agnostic. It does not deny the existence of mental states, but argues that they are one and the same as neurophysiological processes, and that it is unprofitable to attempt any sort of scientific analysis based on introspective reports. Half a century of behavioral science has progressed on this basis, along with many discoveries in neurobiology in the broadest sense, including ethology. But what was originally an agnostic position tended to drift implicitly into a sort of *de facto* denial that mental states or consciousness exist outside our own species.

It is very easy for scientists to slip into the passive assumption that phenomena with which their customary methods cannot deal effectively are unimportant or even nonexistent. To quote Fouts (1973): "All one needs to do is to look around and *not* see something and then conclude that the thing that was not seen in a particular species is totally absent in that species." Here I should also like to follow the example of Holloway (1974) in quoting Daniel Yankelovich ("Smith," 1972): "The first step is to measure whatever can be easily measured. This is okay as far as it goes. The second step is to disregard that which can't be measured or give it an arbitrary quantitative value. This is artificial and misleading. The third step is to presume that what can't be measured easily isn't very important. This is blindness. The fourth step is to say what can't be easily measured really doesn't exist. This is suicide."

Biological evolution is universally accepted by behavioral scientists as historical fact. Animals are used as surrogates or "models" for behavioral investigations on the implicit assumption that principles discovered in this way are applicable to our own species, as discussed in Chapter 10. Certainly this assumption implies qualitative continuity. If, for example, all human learning were believed to be radically different in kind from that available for analysis in other animals, no one would even suggest applying to questions of human education what has been learned by studying rats, pigeons, or monkeys. Yet, when questions of communication and language arise, even hard-nosed behaviorists take for granted a large element of discontinuity. It is indefensibly circular to argue that language is unique to man and, therefore, no matter how complex animal communication turns out to be, it cannot possibly be comparable to human language.

Must we reject evolutionary continuity in order to preserve

our gut feeling of human superiority, as proposed by Adler (1967), Critchley (1960), Langer (1967, 1972), and many others? Or can we be satisifed with a merely quantitative, if enormous, difference between communication behavior in our own and other species? If we insist on a qualitative human-animal distinction in the area of communication behavior, a radical difference in kind in Adler's terms, must we support our insistence by criteria as subjective and difficult to test as those that were rejected by the founders of behaviorism?

The rigid position of the strict behaviorists has been questioned with increasing frequency. For instance, Mowrer (1960b) introduced a chapter entitled "Images, Memory, and Attention (Observing Reactions)" with the remark that these terms have been "and perhaps are still, in some measure *taboo*. Many of us were taught, under pain of banishment from professional psychology, never to use these terms, at least not during 'working hours' . . . such language was deemed completely unsuited to the purposes of science. . . . But it is slightly ironical that those very methods of analysis and research which radical Behavorism introduced are now leading us, ineluctably, back to concepts which Behaviorism was determined to ignore—or even destroy."

In facing squarely the problems of dealing with the possibility that animals have mental experiences, it may be helpful to recognize that our current climate of opinion in the behavioral sciences involves a gradient of acceptability concerning the terms and concepts listed on page 114:

```
O. K.   PATTERN RECOGNITION
   ▲    NEURAL TEMPLATE
   │    SOLLWERT
   │    SEARCH IMAGE
   │    AFFECT
   │    SPONTANEITY
   │    EXPECTANCY
   │    COVERT VERBAL BEHAVIOR
   │    INTERNAL IMAGE
   │    CONCEPT
   │    UNDERSTANDING
   │    INTENTION
   │    FEELING
   │    AWARENESS
   │    MENTAL EXPERIENCE
   │    MIND (MENTAL)
   │    THOUGHT
   │    CHOICE
   ▼    FREE WILL
 TABOO  CONSCIOUSNESS
```

Individual scientists might wish to rearrange some entries in this rank order of orthodoxy, but there is no doubt that the gradient is a significant reflection of the current *Zeitgeist*. Rearranging these terms like playing cards is an entertaining game, but few radical rearrangements would leave the list a plausible one. It is also instructive to ask where one should draw a line to represent the boundary of scientific validity. Very strict behaviorists might stop after Affect, others may venture farther down the list. There are, of course, many philosophers who disagree with positivism, and they feel comfortable with a list extending beyond this one in the direction labeled Taboo (Fodor, 1968*; Feigl et al., 1972*; Polten, 1973*).

Perhaps Jennings and Thorpe have outlined the most reasonable view, considering the limited evidence available: that the gradient is a true continuum without sharp discontinuities. Furthermore, it seems more likely than not that certain animals have mental experiences involving, to varying degrees, the attributes represented crudely by this rank-ordered list of terms.

Many branches of science have made significant and substantial progress by employing postulated entities that could not be observed directly, at least when first developed, but which were inferred from observations of their supposed causes and effects. Gravitation, electric potentials, magnetic fields, atoms, neutrinos, x-rays, chemical bonds, hormones, genes, and nerve impulses are pertinent examples. The impossibility of neatly verifying the existence of mesons or quarks has not inhibited theoretical physicists. Nor has the well-known dilemma concerning the wave and quantal properties of electromagnetic radiation led physicists to stop all investigations of quantum mechanics or particle physics simply because they cannot yet tell us whether light is waves or particles, or explain how it can have the properties of both at the same time. Likewise, paleontologists do their best to make sense out of the fossil record and sketch in evolutionary sequences or unfossilized morphologies without realistic hope of obtaining specific verification within the foreseeable future. Astrophysics is also based on concepts about events and processes immune from direct observation by any methods we can yet imagine.

Investigators of behavior have attempted to formulate comparable explanatory concepts, such as motivation, drives, or Lorenz's specific action potential. But perhaps we have been overlooking more directly pertinent concepts lying close to hand or even closer—inside our own heads. When thinking about Washoe in the act of exchanging information about objects, actions, or desires via manual gestures, or when contemplating Lindauer's swarming bees dancing about the suitability and location of cavities where the swarm might find a new home, I submit that it may actually clarify our thinking to entertain such thoughts as "Washoe *hopes* to go out for a romp, and *intends* to influence her human companions to that end," or "This bee *likes* one cavity better than the other, and *wants* her swarm to occupy the preferred one."

Of course, the use of such terms as *want* or *like* does not explain the basic causes of the observed behavior or of any mental experiences that may accompany it. Nor should the use of these or similar terms be taken to imply identity with any human mental experiences. The degree of similarity or differences would be a stimulating possibility for future investigations. Perhaps this return to a consideration of basic subjective qualities can supply a unifying framework into which many complexities of animal behavior can be fitted. To paraphrase Boyle (1971), perhaps we can understand how, and to what extent, animals make sense of the flow of events of which their behavior forms a part.

Most people not indoctrinated in the behaviorist tradition take it for granted that animals do have sensations, feelings, and intentions. This intuitive impression is based on our experience with patterns of animal behavior that appear sufficiently analogous to some of our own behavior to permit us to empathize. The dilemma of contemporary behavioral scientists results from our indoctrination that *as scientists* we must put such notions behind us as childish sentimentality unworthy of a rigorous investigator (Hebb, 1974). Yet the behavioristic and reductionistic parsimony typified by Watson and Loeb may have led us down a sort of blind alley, at the end of which we find ourselves defending to the last, at least by implication, a denial of mental experience to animals, a denial which we cannot justify on any explicit basis except the presumed absence of communication with conscious intent.

As learnedly discussed by Malcolm (1973*), even Descartes, the fountainhead of the philosophical view that animals are merely machines, admitted that they could feel pain or pleasure and express passions. Yet many behaviorists believe that it makes no difference whether one thinks in terms of possible mental experiences or simply in terms of stimuli and responses, however complex. But the same argument can be applied to people. Inasmuch as we have only indirect evidence about *their* mental experiences, we may logically question whether they really exist. But if questions are raised by others about the reality of one's own subjective feelings, who is likely to fall back on a negative, or even an agnostic, response?

8

Objections
and their Limitations

To reopen the questions discussed in this book that have long seemed irrelevant to twentieth-century behavioral science is disturbing in many ways. Many behavioral scientists would clearly prefer to pack all of these notions back into the secure Pandora's Box where they have quietly rested for so many years (reviewed by Lorenz, 1958, by Klopfer and Hailman, 1967, by Klein, 1970, by Stenhouse, 1973, and by Schultz, 1975). Others appear simply to prefer statements of faith that man is radically different in kind from all other animals and, furthermore, is intrinsically superior, not only mentally but in fundamental moral values. These deep-seated objections deserve careful attention, for it is surely no accident that they are so widely and strongly felt.

The Behavioristic Objection

Virtually all comparative psychologists—and ethologists, as well—are at least *de facto* behaviorists in the sense that they concern themselves only with observable behavior and shun any involvement with possible subjective qualities or mental experiences (Lashley, 1949*). Watson defined behaviorism operationally: "Behaviorism . . . holds that the subject matter of human psychology *is the behavior of the human being*. Behaviorism claims that consciousness is neither a definite nor a usable concept. . . . Its closest scientific companion is physiology. . . . It is different from physiology only in the grouping of its problems, not in fundamental or in central viewpoint." Biologists concerned with animal behavior have adhered, with very few exceptions, to the comparable tenets of Jacques Loeb (1900, 1912, 1916) and C. Lloyd Morgan (1894). It may be well to repeat Morgan's canon here: "In no case may we interpret an action as the outcome of the exercise of a higher psychical faculty, if it can be interpreted as the outcome of the exercise of one which stands lower in the psychological scale." This has been widely interpreted as requiring that complex functions should not be postulated if a simpler explanation will suffice. That is the widely accepted principle of parsimony: given a choice of two or more plausible explanations, the simplest is preferred. Any suggestion that animals might be aware of the flow of events with which their behavior interacts has long been rejected as unparsimonious. Animal behavior, and even human behavior, seem simpler and more manageable if we persuade ourselves that subjective mental experience is of no consequence. As a result, only the most physiological explanations have customarily been recognized as worthy of scientific consideration.

The strict behaviorist believes it is operationally meaningless, and hence foolishly unscientific, to consider even human mental experiences, because they cannot be observed directly and because verbal reports about them are inconsistent and impossible to verify by any other means. S. R. Brown (1972) has clearly stated this dilemma. "Human subjectivity is a phenomenon that has both interested and eluded social scientists for some time. . . . Perhaps beginning with Watson, the behaviorists rejected introspection and romantic mentalisms, and they did so for sound scientific reasons: As scientists, they knew no way of dealing with these mental goings-on. The rejection, however, was regarded at the time (except, perhaps, in the case of Watson himself) as temporary, pending the development of instrumentation capable of dealing with what previously had proved elusive. The second generation behaviorist forgot the original reasons for the rejections, remembering only the act of rejection itself; the third generation behaviorist was even at a greater disadvantage, being unaware anything was even forgotten."

When Lashley (1923) asserted that "The supposedly unique facts of consciousness do not exist," he was rebelling against schools of psychology which held that mental qualities were different in kind from physiological processes, and hence could never, in principle, be explained in physicochemical terms. Lashley was attacking what he saw as the subjectivists' belief in a separate psychic world, "a unique mode of existence not definable in objective terms." He viewed introspection as "an example of the pathology of scientific method." But he also conceded that "there can be no valid objection by the behaviorist to the introspective method so long as no claim is made that the method reveals something beside bodily activity."

Hebb (1974) argued vehemently: "Subjective science? There isn't such a thing. Introspectionism is a dead duck." Yet later in the same article he asserts that "Psychology is about the mind; the central issue, the great mystery, the toughest problem of all." But it is not clear how a central probem can be solved by ignoring most or all of the available evidence. Skinner (1957) was well aware of the danger that important problems might be ignored merely because they are difficult to study. He attempted to deal with what are generally called mental processes, or thinking, while maintaining a consistent behavioristic position. This he did by treating thinking as *covert verbal behavior*, in which the speaker and listener are the same person. This position equates thinking with a sort of talking to oneself, which is almost as unobservable as the mental concepts which Watson and the original behaviorists rejected as unworthy of scientific mention. But Skinner saw an important difference, in that such covert verbal behavior can be influenced by prior and subsequent events, especially by reinforcement, and thus, like other behavior, its properties can be deduced from an analysis of its antecedent causes and subsequent results. Because Skinner's behaviorism was constrained to deal only with input-out-put relationships and methods of reinforcement, it was possible, by stretching its definitions, to include even unobservable, covert verbalizing; at least, Skinner felt justified in doing so.

Skinner's definition of thinking as covert verbal behavior can reasonably be extended to include not only the literal exchange of words and sentences, but also the many other forms of communication used by human beings, such as signs, gestures, mathematical symbols, and the whole nexus of nonverbal communication. Lashley (1923, p. 342) anticipated important elements of Skinner's position, including this broadening of its scope: "The relation of any integration to the speech and gestural mechanisms is of prime importance for its 'conscious aspects.' Not only is the single certain evidence of consciousness in another person the existence of consistent, rational expressive movements. . . . The core of 'conscious' integration is the verbogestural coordination." New discoveries have made it reasonable to include consistently communicating animals within Lashley's definition. Extensions of this approach are summarized in McGuigan and Schoonover (1973).

To include both nonverbal and verbal communication in a behavioristic definition of thinking, along the lines advocated by Lashley and Skinner, intuitively seems even less an extension of the behaviorist position than is the redefinition of thinking as covert exchange of words and sentences. To accept verbal, but not nonverbal, communication into the covert fold would scarcely be parsimonious. If we extend Skinner's concept of covert verbal behavior to covert communication behavior internally within the human brain, we may then ask: Why not also within the chimpanzee brain? Washoe and the other chimpanzees that have learned to use gestures which serve many of the simpler functions of words in human speech can be assumed to manipulate covertly whatever communication systems they use overtly. Can birds sing to themselves covertly, or do vervet monkeys give inaudible alarm calls when uncertain whether a distant speck in the sky is a dangerous eagle? Can weaver ants give internalized recruiting gestures and bees dance to themselves when motivation for communication behavior is present but, for some reason, the overt behavior is not possible? Some readers may feel they have been led astray and that something must be wrong, because we have reached the unwelcome conclusion that animals can think. But where is the error? Is it in the argument that if words can be employed covertly, so can nonverbal means of communication? Or in carrying over the notion of covert communication from people to chimpanzees? Is this notion acceptable in apes, but not in birds or bees, and, if so, where and how is a line to be drawn?

One objection that can be anticipated is the assumption that animals always express their motivational states immediately, whereas men can inhibit the expression and yet retain the internal communication behavior. But this seems highly unlikely as a general rule, for in many cases animals clearly retain the memory of some relationship about which they communicate only when circumstances are appropriate. One example from honeybees may suffice to make this point. If bees from a colony with a severe need for food have been foraging at a certain food source and dancing actively about its location and quality right up to sundown, they will ordinarily do no dancing during the night. But as soon as morning comes they will fly out to the same source, taking now a very different direction relative to the sun (Frisch, 1967). One could explain this behavior by postulating, quite reasonably, that they had learned other cues, such as landmarks, that led them to the food. But under some circumstances, such bees can be stimulated to dance during the night if appropriate lighting conditions are provided. They dance with the waggle runs oriented at angles to gravity intermediate between the directions indicated in the evening and morning, with the difference roughly proportional to the time that has elapsed since sunset. Not only does this demonstrate the existence of an endogenous biological clock and continuous correction for the passage of time; it also suggests that the memory of food location and the motivation to communicate about it remain present in a latent, covert state somewhere within the bees' nervous systems. Covert communication behavior can be inferred on the basis of these experiments, and its properties deduced from the effects of prior causes and subsequent results, just as Skinner attempts to do with covert human verbal behavior considered as an objective definition of thinking.

The assumption that animals always respond in a rigid, machineline fashion to immediate stimuli is widespread. However, in laboratory experiments, animals can be trained to respond not immediately, but after waiting for varying intervals. In general, animals we consider "higher" can learn to make such responses after holding themselves in check for longer intervals of delay. Laboratory experiments have been devised which greatly reduce the likelihood that the animal is simply rehearsing the learned response. Psychologists have struggled to explain what keeps an animal ready to respond after appropriate delay by calling it "bridging." But this seems a major problem only if one's thinking about animal behavior is constrained within the narrow limits of conventional, behavioristic learning theory. If we assume that the animal simply understands what it has learned, the delayed responses cease to be especially puzzling. Perhaps postulating simple thoughts in the minds of animals may result in more parsimonious, as well as more nearly correct, explanations.

It is reasonable to ask whether wild animals ever show comparable delayed reactions under natural conditions. Very few examples of this have been demonstrated convincingly, perhaps because of the difficulty of identifying the actual stimulus if the response to it occurs only after a long and unknown lapse of time. Many insects and birds learn to come to food sources only at the particular times of day when they are available, for instance, flowers that provide accessible nectar only at about the same hour of the day or night. One can interpret such behavior as a response delayed for roughly 24 hours after the last reinforcement, but most behavioral scientists would tend to postulate instead some sort of endogenous circadian periodicity with which the availability of food has been associated. Another common pattern of behavior that involves long delays between two or more actions is the storage of food and its retrieval after weeks or even months. The squirrel that has buried a nut, the beaver that has accumulated branches in an underwater food pile, or any other animal that stores food for long periods can be viewed as being stimulated by the food and its location when storing it (when hunger is clearly satisifed for the moment) and then delaying return until a later time, when the animal is hungry and unable to find other food. Neither of these categories of delayed behavior matches the experimentally arranged delayed-response test, but the lack of naturally occurring analogues may reflect not so much an inability of wild animals to delay their responses, as a rarity of situations in which such delayed responses would be useful.

The opinions of Mowrer (1960b), quoted in the previous Chapter, indicate that many psychologists have been relaxing the rigid strictures of Watson's original behaviorism. An agnostic reservation of judgment is clearly the soundest position at present. But it should be an open-minded agnosticism, which recognizes the possibility that answers not yet available may be obtained from future investigation of questions about mental experiences in animals.

Semantic Behaviorism

Throughout the long period when experimental psychology was dominated by behaviorism, many ingenious and effective experiments were devised to analyze animal learning and problem-solving. Although mentalistic terms and concepts were scrupulously avoided, it seems clear in retrospect that a large fraction of the scientific interest has actually been directed toward the possibility of animal thinking. This is particularly true when the more complex types of discrimination learning have been under study, and when experiments have suggested that animals may employ concepts of various kinds or understand simple, but somewhat abstract, relationships. As animal learning and problem-solving were gradually found to be more complex and versatile, the taboos of behaviorism have continued to receive lip service, but with diminishing adherence to the strict tenets of the early behaviorists. Experiments and results are described in behavioristic terms, even though they strongly suggest that simple thinking is going on in the brains of the animals under study. We have thus reached a stage at which a sort of *semantic behaviorism* results in complex circumlocutions and confusing euphemisms. These are often far less parsimonious than frankly calling a spade a spade or a thought a thought.

Köhler (1925) and Yerkes and Yerkes (1929) demonstrated that the Great Apes could learn to solve a wide variety of novel problems not likely to have been encountered under natural conditions. Many more recent experiments with monkeys, cats, rats, and several species of birds have also yielded results which can easily be interpreted as evidence of simple kinds of conscious thinking (Mackintosh, 1974; Hulse et al., 1978). To be sure, behaviorists can always respond to such an interpretation by arguing that appropriate and versatile learned behavior or problem-solving might be accomplished without any conscious awareness. But these arguments are essentially cautionary, and not decisive.

The Anthropomorphic Objection

A common objection to the notion that animals have mental experiences is the charge that such thinking is anthropomorphic, that is, that it requires ascribing human thoughts to other species. But it is actually no more anthropomorphic, strictly speaking, to postulate mental experiences in another species than to compare its bony structure, nervous system, or antibodies with our own. There is a serious danger of circular reasoning in basing a denial that animals can have mental experiences on the mere assertion that this suggestion is anthropomorphic. The charge carries weight only if one assumes in advance that animals do *not* have such experiences, and thus the accusation merely reiterates the original assumption.

The prevailing view implies that only our species can have any sort of conscious awareness or that, should animals have mental experiences, they must be identical with ours, since there can be no other kind. This conceit is truly anthropomorphic, because it assumes a species monopoly of an important quality. It resembles, in many ways, the pre-Copernican certainty that the earth must lie at the center of the universe. As I have mentioned, there is no reason to believe that any mental experiences animals may have must be identical to our own. Indeed, there is no reason to suppose that they are absolutely identical throughout *Homo sapiens*—for instance, between men and women, or children and adults. As Wittgenstein (1953) puts it, "one human being can be a complete enigma to another. We learn this when we come into a strange country with entirely strange traditions; and, what is more, even given a mastery of the country's language. We do not *understand* the people. (And not because of not knowing what they are saying to themselves.) We cannot find our feet with them."

Subjective qualities and mental experiences have remained largely untouched by the Darwinian revolution, primarily for lack of effective methods for detecting them reliably in other species, let alone analyzing them by scientific methods. But, in our present state of ignorance, we certainly cannot exclude the possibility that mental experiences, like other attributes of animals and men, exhibit continuity of variation and are not typologically discrete, all-or-nothing qualities totally restricted to a single species.

When we attempt to imagine what an animal may be thinking in a given situation, we are obliged to describe in words (or perhaps in human gestures or expressions) what we suggest the animal may feel. Critics object that the mere use of human language introduces a crippling error, because the animal does not use words. But this may be no more than a methodological problem that could be overcome by careful procedures. A beginning has been made in recent experiments by Premack and

Woodruff (1978), in which chimpanzees are asked to select one of several photographs that represent a particular pattern of thought which the animal may be experiencing. In "blind" tests, the experimenter presents pictures to a chimpanzee in a manner that allows the experimenter no opportunity to see the picture himself. In some cases, at least, the trained ape gives the appropriate gestural sign to indicate recognition of the correct solution to a particlular problem. Although much remains to be done before we can be sure just what the signs or "words" actually mean to the apes who have learned to use them, the signing behavior certainly offers some indications of what the chimpanzee may be thinking about. This is particularly true when signing occurs spontaneously and in the absence of the objects designated by the signs.

The problems raised by using human words to describe postulated thoughts of animals would be critical only if human thoughts were absolutely identical with words and sentences, if they never took a form not conveyed precisely by a verbal description, and if no other kind of thought existed. It seems highly unlikely that any of these conditions are absolutely necessary, even for our own species (Hutchinson, 1976). They imply, for example, that new experiences could never occur in the absence of appropriate words already shared by a group of people speaking the same language. This would be a serious limitation of the creativity held by Chomsky and many others to be such an important attribute of human language. Granting that words express thoughts more or less imperfectly, the residue of properties not absolutely coextensive with words might be supplied by covert nonverbal communication in the sense discussed earlier. This possibility is more difficult to evaluate and, by extending the concept of covert nonverbal communication, one might well be able to include all vague feelings, impulses, emotions, and so forth. But whether this can be done in any convincing fashion is not critical to the present discussion.

A further objection that may confidently be anticipated is that interspecific communication could occur only between very closely related species, say man and chimpanzee, but not man and dog and still less man and bird, whereas man-to-bee exchanges of information would be quite unthinkable because of the remoteness of any common ancestors. Many will agree with Wittgenstein (1953) that "If a lion could talk we could not understand him." Thus, Nagel (1974) has argued that we can never know what it is like to be a bat or, presumably, any other species not very closely related to our own. This widely held view contains the implicit assumption that mental experiences vary rapidly with branching evolutionary lines of descent. But the mere possibility that mental experiences might exist in animals has been so thoroughly ignored by behavioral scientists that we naturally have no data whatever about how such experiences may vary between species. It is certainly possible to imagine at least some of the experiences a distantly related animal might have. The problem is how we might test such speculations, and in Chapter 10 I will try to suggest some promising approaches to this thorny problem.

An equally plausible hypothesis is that, as mental experiences are directly linked to neurophysiological processes—or absolutely identical with them, according to the strict behaviorists—our best evidence by which to compare them across species stems from comparative neurophysiology. To the extent that basic properties of neurons, synapses, and neuroendocrine mechanisms are similar, we might expect to find comparably similar mental experiences. It is well known that basic neurophysiological functions are very similar indeed in all multicellular animals. On this basis, we might be justified in turning the original argument of the strict behaviorists completely upside down. Because neurophysiological mechanisms appear to be very similar in men and bees, the mental experiences resulting from their operation must, according to this line of reasoning, be equally similar. If this seems an embarassing conclusion, we can try to escape from it by postulating that neurophysiology is seriously incomplete, having failed, so far, to locate those functions that differ so widely between taxonomic groups that they generate incomprehensibly divergent mental experiences. One alternative is to postulate some special form of human uniqueness, not demonstrable objectively—an unparsimonious procedure, to say the least. Another is to postulate that human uniqueness lies in the patterns or "programs" into which neurophysiological components are organized.

"It's Only Problem-Solving"

Particularly in discussions of the signing apes, and of the experiments by Premack and the Rumbaughs, in which captive chimpanzees have been trained to use various types of artificial symbol systems, one often encounters the objection that all this seemingly languagelike behavior can be explained more parsimoniously by assuming that the animals have simply learned to solve specific problems. This viewpoint goes on to interpret the evidence in the following general way: the trained apes have learned to do certain things, whether these be making certain gestures, pressing particular keys, or manipulating plastic objects, in order to obtain desired rewards. These rewards may be food, or obviously pleasurable social interactions with human companions. To maintain the distinction between this interpretation and the ascription to languagelike behavior entails the assumption that human language is fundamentally different from problem-solving. Of course, our language is often used to solve problems or obtain something we desire, and, in the very early stages of language-learning, young children may well learn only that a certain sound yields a coveted goody. But to whatever extent it may begin as problem-solving, human language obviously comes later to be used in a much more complicated and more versatile manner.

In a short note often cited in support of this interpretation, the late E. H. Lenneberg is quoted (1975) as having "trained normal high school students with the procedures described by Premack, replicating Premack's study as literally as possible. Two human subjects were quickly able to obtain considerably lower error scores than those reported for the chimpanzee. However, they were unable to translate correctly a single sentence, completed by them, into English. In fact, they did not understand that there was any correspondence between the plastic symbols and language; instead, they were under the impression that their task was to solve puzzles. Further, they tended to forget the solution to a task almost as soon as they were confronted with new tasks, so that, in the end, they were not able to answer questions (i.e., give the correct response by means of symbols) randomly chosen from their total repertoire."

Unfortunately, this brief note leaves several key questions unanswered. Which of Premack's procedures were replicated? And what instructions were given to the human subjects? In the absence of any statement to the contrary, one assumes they were *not* told that they were to learn a simple form of language, and hence it may not have been reasonable to suppose that they could recognize what they had learned as forming the basis for a communication system. It is also puzzling that normal high school students were unable to recall problems learned earlier in their tests; this suggests that they were not led to expect that such recall would be expected of them. If so, this would also tend strongly to discourage any use of the learned procedures for communication. Lenneberg certainly raised a significant question, and his general approach should be refined and extended, but this brief published account is inconclusive.

Human language is commonly distinguished from even the most complex kind of problem-solving by its flexibility and creative versatility. It was largely on this sort of basis that Chomsky (1959) criticized the behavioristic analysis of language so effectively that very few psycholinguists adhere to Skinner's interpretation of human language and thinking. But suppose an animal learns not only to solve a specific problem but to generalize from that task to other more or less similar problems? How do we draw the line between true language, on the one hand, and increasingly adaptable, perhaps even creative, problem-solving? Spontaneity and the coining and use of new communicative signals or, better still, rule-guided and meaningful combinations of such signals, are characteristic of human language. But available evidence leaves much doubt as to whether these attributes are present in the seemingly languagelike behavior taught to apes. Yet animals sometimes use communicative behavior to solve newly arisen problems by coordinated group action.

The disparaging interpretation "It's only problem-solving" is actually accompanied by several other implicit assumptions, such as fixity of the animal's response to individual problems and lack of generalization to other problems. The rarity or absence of reports of spontaneously creative communication by signing apes is perhaps a more serious reason for doubting that they are really expressing thoughts, rather than performing elaborate behavior patterns they have learned without even the most rudimentary understanding. But, as also mentioned above, it is not clear whether the experimental situations used to study these signing apes are well suited for learning to what extent spontaneously creative use of learned communicative gestures does actually occur.

The Dangers of Relaxing Critical Standards

Opening our minds to consider the possibility that at least some animals may have mental experiences is only a first step, though a crucial one that requires a significant departure from the current *Zeitgeist* of the behavioral sciences. Having taken this step, one is tempted to plunge at once into the very different process of inferring particular mental experiences in specific animals. This second step is as hazardous as it is seductive. It is so easy to guess about the processes occurring in the brains of another species, on the basis of its observed behavior, that we can easily forget how many such confident conjectures have come to seem implausible as more has been learned about the natural behavior of the animals concerned. Careful ethological analysis of preceding and subsequent behavior can provide some indications of the possible mental experiences that may exist within a given animal under particular circumstances. But we would be on much firmer ground if we could obtain more direct access to whatever events may be taking place inside the animal's head. Some possible approaches to this difficult task are suggested in Chapter 10.

It now seems timely to relax to some degree the reductionist and behaviorist reactions that led to the inhibitions discussed above. George Miller (1962) has pointed out that Morgan's canon was proposed, not with any intention of avoiding introspection, nor with any doubt that mental experiences exist in both men and animals. Rather, as Miller puts it, "all that Morgan hoped for were a few reasonable rules for playing the anthropomorphic game." The view that it is anthropomorphic to postulate any sort of mental experiences in animals may have resulted from a confusion of scientific caution and parsimony with unscientific feelings of human superiority.

The Clever Hans Objection

There is no doubt that enthusiastic observers of animals are constantly in danger of interpreting their behavior in more complex terms than is necessary or correct. Clever Hans is an outstanding example, as discussed in Chapter 4 (Pfungst, 1911). It seems inherently reasonable that the ability of a horse to notice when a human experimenter had stopped making small counting movements and was waiting expectantly for the horse to end its tapping, or showed other signs of expecting something to happen, is a simpler and far more convincing interpretation than the conclusion that this or other horses could really carry out complex arithmetical manipulations, such as multiplication and division of numbers written on a blackboard.

Recently, Sebeok and Umiker-Sebeok (1980) have argued that many or all of the claims that apes have learned some elements of human language suffer from Clever Hans errors or comparable misinterpretations based on an uncritical desire to find evidence of animal thinking. Similar criticisms have been applied to reports of what appear to be complex or intentional communication in other animals. But, while this hazard must always be kept in mind and guarded against, it is difficult to see how the communication discussed in Chapters 3 and 4 can be merely figments of wishful imagination on the part of the scientists who have analyzed them. To recognize that any mental experiences animals may have need not be identical, or even necessarily similar, to those of a man under comparable conditions, opens up a wider range of potential interpretation but, at the same time, makes it more difficult to gather convincing data. In Chapter 10, I suggest possible solutions to this dilemma. For the time being, it must suffice to emphasize that the problem is serious. Cautious treading of a middle ground is clearly called for to avoid both of two obviously fallacious extremes: (1) the postulation of complex mental activities (such as horses capable of long-division) when simpler ones are consistent with the observed behavior of the animal and the observed responses of conspecifics to its communication signals; and (2) the conventional reductionist position that animals have no mental experiences at all, or that any they may have are hopelessly inaccessible to our investigation.

As we begin to study the possibility that mental experiences play a significant role in animal behavior, we should be warned by the unjustified definiteness of the quotations cited in previous chapters.

The "So What?" Objection

One vestige of strict behaviorism takes the form of asking disparagingly what difference it would make in our ideas about animal behavior, or our investigations of it, if we *did* postulate mental experiences in the animals under study. Surprisingly enough, many ethologists have taken over this aspect of behaviorism. To be sure, some investigations and conclusions will depend very little on the distinction between a Cartesian concept of animals as purely deterministic, unconscious machines and a broader view, which accepts the possibility that conscious intentions and mental experiences may be present. One can derive predictive generalizations by considering only one aspect of the phenomena under study. For example, one can predict the caloric value of plant or animal tissues by burning them in bomb calorimeters, and such experiments have demonstrated the important fact that the energy released by oxidation of living tissue can be predicted by the same chemical principles that govern the heat of combustion of inorganic substances. It is completely irrelevant to such investigations whether the living tissues come from animals, plants, brains, leaves, roots, ovaries, or fingernails. If one is interested only in heat of combustion, one can ignore these distinctions.

This example is a reduction to the absurd, but when we are dealing with phenomena as complex and subtle as those of behavior and social communication, it is prudent to keep an open mind concerning which attributes of the systems under study may be important. If one is interested only in relatively straightforward predictions of simply described aspects of behavior, the Skinnerian viewpoint may be wholly sufficient.

On the other hand, there is considerable reason to believe that in at least one species, *Homo sapiens*, mental experiences play a significant, though not all-encompassing, role in the regulation of behavior. Accepting the reality of our evolutionary relationship to other species of animals, it is unparsimonious to assume a rigid dichotomy of interpretation which insists that mental experiences have some effect on the behavior of one species of animal but none at all on any other. It would be absurd to deny that mental experiences are important components in human behavior and human affairs in general. To the extent that animals have them, mental experiences may also be significant in their activities. It is obvious that one could not understand human beings as well, or predict their behavior as accurately, without taking some account of their awareness and intentions. The same consideration applies to other species, insofar as mental experiences are significant in their behavior.

An Obsolete Strait Jacket

It is important at this point to recognize that ignoring the possible existence of mental experiences and conscious intent in animals may have held back our scientific progress in this important field, as anticipated by Jennings (1933). It seems possible that the variety of communications conveyed by the dances of bees might have been discovered by Frisch in the 1920s, if anything like complex communication among insects had not been so utterly unthinkable. This question leads to another: What are we *now* overlooking, as a result of comparable restrictions imposed on the questions we ask, by our basic viewpoint about the nature of animal and human behavior?

The extensive data gathered by behavioral scientists often seem to be filled with confusing contradictions; perhaps unifying concepts might be discerned more easily if mental experiences were included within the scope of our hypotheses and explanations. If so, their recognition would indeed make a difference. The first example that comes to mind is the injury-feigning, or predator-distraction, displays of certain birds. Ordinarily this type of behavior results when an adult bird incubating eggs or caring for its young is disturbed by the approach of a man or other potential predator. The bird acts as though it is injured, flutters in a manner suggesting that it has a broken wing, and tends to move slowly away from the nest or young. Often the potential predator does, in fact, follow the "injured" bird and, after an appreciable distance separates the parent from its young, the bird suddenly recovers its normal capabilities and flies away (Brown, 1962; Gramza, 1967; Wilson, 1975; Skutch, 1976).

During the past half-century, ornithologists and ethologists have gone to great lengths to deny that such birds could have any conscious intention to lead the predator away from its offspring. For example, in a general review of predator-distraction displays, Armstrong (1949) felt it was important "to have available a series of terms that are devoid of disputable cognitive implications." The term injury-simulation was preferred to injury-feigning because it "carries the sense of deliberate intention to deceive rather less than 'feign' and is, therefore, preferable." Armstrong suggested that "distraction displays have arisen through the 'displacement' of components from other behavior contexts, particularly threat and epigamic display, which have become ritualized into new behaviour-patterns with survival value." This strongly implies that a ritualization of displays which originally served other purposes is an adequate explanation and, furthermore, one which removes any need to be concerned about whether the birds are consciously aware of the probable results of this distinctive display. It is quite possible that both interpretations are correct; that an animal can consciously employ ritualized display patterns which have the evolutionary origin suggested by Armstrong and others. No adequate data are available to resolve such differences, but they are worth discussing.

A second example in which the possible presence of mental images may help make sense out of behavior patterns that otherwise require complex explanations are the elaborate structures built by certain species of birds as part of their mating displays. Some of the most extreme examples are the bowers of the bower-birds and their relatives in Australia, New Guinea, and adjacent islands (Marshall, 1954). In some species, the bowers are remarkably complex structures built from twigs, grass, and other vegetation in small areas which the bird has cleared. They are often decorated with conspicuous objects, such as fruits, flowers, fungi, and occasionally with silver coins, jewelry, or even automobile keys. The males display at these bowers and females are attracted by the displays, which presumably play a significant role in mate selection. Marshall and others are vigorous in their denials of any interpretation of bower-building that implies conscious intent on the part of the bird. "These and other singular attributes have caused a voluminous popular literature to spring up about the family. Much of this is nonsense. Most of it has been marred by anthropomorphic generalization, and all of it is unsupported by experimental evidence." Marshall is emphatic that his aim is "to describe these and associated phenomena in terms of animal rather than human behaviour. . . . These complex and remarkable phenomena are probably expressions of innate behaviour patterns that are annually called into play by the secretion of sex hormones. . . . The theories of Australian naturalists that bower-birds are especially intelligent and that their display activities are largely 'relaxative', consciously aesthetic, and unconnected with the sexual drive are rejected, though of course it is not suggested that the birds do not enjoy the fantastic activities that they perform." Hartshorne (1973) is likewise inclined to believe that, in song birds, "Song expresses feeling, according to principles partly common to the higher animals. That a bird sings 'because it is happy' is not entirely foolish."

In the final chapter of his book, Marshall returns to questions of intelligence and estheticism in bower-birds. "While I have ascribed a utilitarian basis for each of the behavioral phenomena discussed, I see no reason, provisionally, to deny that bower-birds possess an aesthetic sense although, it must be emphasized, we have as yet no concrete proof that such is the case. Some bower-birds certainly select for their displays objects that are beautiful to *us*. Further, they discard flowers when they fade, fruit when it decays, and feathers when they become bedraggled and discoloured. But, it must be remembered, however beautiful such articles may be, they are still probably selected compulsively in obedience to the birds' heredity and physiology." Marshall thus assumes that bower-birds behave "compulsively" and he seems to imply that although they may enjoy their "fantastic activities," they do so without conscious awareness of the results of their behavior.

These bowers are extreme cases out of a huge spectrum of behavior in which animals alter their immediate environments by constructing shelters or arenas used for mating displays. We have become so accustomed to concentrating on functional and adaptive aspects of these behavior patterns that we have neglected even to ask whether the animals have any awareness of the probable consequences of their behavior. We do not yet have available direct evidence indicating whether they do or do not intend to influence perceived future events, but it may be a serious limitation in our thinking to assume *a priori* that no such awareness can possibly exist.

The possibility that animals have subjective feelings of various kinds has been ignored as studiously as has the possibility that they might have intentions or mental images. Behavior such as building and decorating bowers or feigning injury raises questions about both the feelings and the mental images that might precede or accompany such behavior. As discussed by Lorenz (1963), the subjective feelings of animals, while even more difficult to study than any mental images they may have, could well be of equal or greater importance. Indeed, most scientists, like other thoughtful people, have little difficulty in accepting the notion that injured animals feel a sort of pain or starved animals a kind of hunger akin to the comparable human feelings. The detailed analysis of animal feelings is another important challenge for ethologists, but one that lies outside the scope of this book.

Perhaps animals perform some of the behavior patterns we observe because they enjoy the resulting experience. For instance, herring gulls often soar for hours back and forth over a particular area where there is an obstruction updraft, with no realistic prospects for food and no evident social function. Such behavior may be adaptively neutral, or virtually so, but may result in a satisfying feeling on the birds' part. One can even postulate that pleasant feelings which result when a physiological capacity is exercised are in themselves adaptive. Even though we cannot yet formulate such concepts as pleasure at all adequately in terms of physiological or biochemical correlates, this does not seem to me to be a sufficient reason for avoiding the concepts themselves as though they were a dangerous plague.

To state that animals do something because they enjoy it is often criticized as tautological. Such criticism argues that the postulation of enjoyment adds nothing to the simple statement that the animal performs the behavior pattern in question. The critic may also go on to object that stating that the animal enjoys a given activity brings one no closer to an ultimate explanation of why it does so. The second objection is valid only against the claim that by postulating enjoyment one has produced an ultimate explanation of the cause for the behavior or the mental experience. But to recognize the probable reality of some attribute is not at all the same as asserting that it constitutes a complete causal explanation. It seems likely that some deep-seated reluctance to think about animal awareness underlies this type of objection to considering, in even the most tentative fashion, that animals may have mental experiences.

I have raised questions that neither we nor our descendants may be able to answer in the next century. Yet scientific progress clearly requires the formulation of alternate hypotheses before the most appropriate questions can even be asked. The customary approach to animal behavior tends to rule out, in advance of any investigation, the possibility that a system of animal communication may be more complex and subtle than is demonstrated by the data immediately at hand, still less that it may involve conscious intention. If the communication system is variable and conveys many fine distinctions, such complications can easily be dismissed as "noise" of no significance. If the animal were to create new messages in an attempt to express a novel thought, even a very simple one, this too could easily be dismissed as noise, because the thinking of the experimenter and the resulting nature of his experiments have, so to speak, too coarse a grain.

9

The Adaptive Value
of Conscious Awareness

That social communication is adaptively valuable to some species of animals is demonstrated by the "cost" of anatomically growing, physiologically maintaining, and behaviorally displaying structures that have communication as their principal or, sometimes, their only known function. An extreme, but not unique, case is the one greatly enlarged claw of male fiddler crabs. The claw constitutes a third or more of the body weight in certain species. The structure is very rarely, if ever, used for anything but social communication—chiefly ritualized aggression between males, and courtship (Crane, 1975). Fiddler crabs and many other animals need more food and are more vulnerable to predation than would otherwise be the case, because of conspicuous structures or conspicuous behavior involved in social communication.

Even more commonly, larger amounts of time and energy are consumed in intermale aggression or courtship behavior than would seem at all necessary for the simple requirement of bringing together males and females ready to mate. These costs seem large compared to those proposed by evolutionary biologists to account for the evolution of morphological characters. For example, the trend for many species of birds and mammals to be larger and to have shorter extremities at higher latitudes is usually explained by postulating that the relatively slight decrease in surface-to-volume ratio reduces heat loss and thus conserves metabolic energy. While such things are difficult to measure, it seems likely that the added cost of competitive group displays at "leks" of grouse and other birds far exceed these differences in heat loss as a function of body surface. Hence, according to the basic axioms of evolutionary biology, such social displays must have been favored by some selective advantage great enough to outweigh their cost. Wilson (1975) discusses these questions with many specific examples.

Communication behavior is probably most likely to resemble human language in species whose social behavior involves a high degree of interdependence, so that it is adaptively advantageous to have an efficient means of communication between individuals to coordinate their activities. But, as social communication is by no means limited to men and honeybees, versatile signaling systems should be advantageous to many species.

Humphrey (1979) and Crook (1980) have elaborated the argument that in human evolution the development of interdependent societies made it adaptively advantageous to recognize other members of one's group as individuals, and to react appropriately to their individual attributes and behavioral idiosyncrasies. They believe that this, in turn, led to the development of symbolic language and the kind of thinking which they and others feel to be possible only with the aid of language. These considerations have so far been applied only to the evolutionary history by which our own species developed from other primates. But the general argument is equally applicable to other social animals.

However vulnerable modern civilized men might seem if forced to live in isolation, under favorable climatic conditions and where food supplies are relatively abundant, some of us would succeed in surviving and reproducing without any of the artifacts to which we have become so accustomed. Yet such a possibility is almost unthinkable in the case of the highly evolved social insects. Their coordinated group activities are so dependent upon effective communication that the adaptiveness of such communicative behavior is even more overwhelmingly evident than in the case of our own ancestors. Individual recognition is felt to be impossible for social insects, but it is not clear whether the kinds of data so far available would reveal individual recognition if it did occur. One can (and perhaps Humphrey and Crook would) accept this argument, but insist that the social communication of insects results entirely from evolutionary selection operating on unthinking automata. But, as discussed in Chapter 5, this belief must rest on other arguments than the adaptive value of symbolic communication for highly interdependent social creatures.

In a wide-ranging and stimulating discussion of the evolution of mental processes, Julian Huxley and Nikolaas Tinbergen expressed a fundamental disagreement concerning the likelihood than animals have subjective mental experiences (Tax and Callender, 1960, pp. 175–206 and 267). Huxley held that they probably do, and that the question is a valid one, open to scientific investigation. Tinbergen argued the contrary position that we have no basis for inferring subjective experiences in other species. During this discussion, Huxley was asked whether conscious awareness is adaptive in the sense that this term is used by evolutionary biologists; that is, whether it has a survival value and hence has been favored by natural selection. Huxley was sure that it does, but his reasons were not stated in any detail in that symposium. Because of its importance, I should like to take up this question where Huxley left off and present arguments that awareness is indeed adaptive.

In strictly operational terms, awareness can be considered as readiness to respond to certain patterns of stimulation. Because responsiveness and awareness are not the same thing, behavioral evidence is indirect and may be unduly limited, or even misleading. For instance, an animal with only one operative sensory channel—an electric fish in muddy water, for example—could be subjectively aware of a simple, but important, communication signal, such as threat of attack conveyed by a change in the frequency of the electric discharges from another electric fish (Hopkins, 1974). Yet, in the absence of other information, such a fish would be operationally indistinguishable from an electronic frequency meter. Must we therefore define awareness in terms of the numbers and complexity of the signal patterns to which an animal is ready to respond, or is the criterion necessarily a subjective one?

An alternate approach is to consider awareness as the existence of internal images available for comparison with current sensory input. This recalls the cybernetic concept of a "Sollwert," the value of a sensory input which the animal tends to keep constant by adjustments of its behavior (Mittelstaedt, 1972). But this concept would have to be extended to include more than one sensory channel. Also related is the notion of a neural template (Marler, 1969). A sufficiently versatile template-matching machine, again in principle, could fulfill the *behavioral* criteria involved here. The psychological concept of a Gestalt is also applicable, at least in part. It is usually defined as a moderately complex pattern recognized from any of several viewpoints or when any of several redundant, overlapping stimulus patterns are perceived. One example is provided by searching images, postulated internal images of something for which the animal searches (Croze, 1970).

The possession of mental images could well confer an important adaptive advantage on an animal by providing a reference pattern against which stimulus patterns can be compared; and it may well be an efficient form of pattern recognition. It is characteristic of much animal, as well as human, behavior that patterns are recognized not as templates so rigid that slight deviations cause the pattern to be rejected, but as multidimensional entities that can be matched by new and slightly different stimulus patterns, as when a familiar object is recognized from a novel angle of view. This ability to abstract the essential qualities of an important object and recognize it, despite various kinds of distortion, is obviously adaptive. Even greater adaptive advantage results when such a mental image also includes time as one of its dimensions, that is, the relationships to past and future events. Mental images with a time dimension would be far more useful than static searching images, because they would allow the animal to adapt its behavior to the probable flow of events, rather than limiting it to separate reactions as successive perceptual pictures of the animal's surroundings present themselves one at a time. Anticipation of future enjoyment of food and mating or fear of injury could certainly be adaptive, by leading to behavior that increases the likelihood of positive reinforcement and decreases the probability of pain or injury.

All these attributes can also be postulated in nonconscious systems, but conscious awareness may be an efficient, and hence adaptive, way in which complex animals cope with changing situations. The matrix of concepts needed to encompass the variety of spatial and temporal patterns successfully dealt with by many animals tends to approach a working definition of conscious awareness. For instance, the image of food within reach might well be coupled with an image of the act of grasping the food, another of swallowing it, or even the image of its pleasant taste. Thus, if the existence of mental images in animals can be accepted as plausible, one need only postulate an appropriate linkage between them to sketch out a working definition of conscious awareness. It may be helpful, and even parsimonious, to assume some limited degree of conscious awareness in animals, rather than postulating cumbersome chains of interacting reflexes and internal states of motivation. Conscious attention to the performance of new and challenging tasks ordinarily improves our performance; perhaps this principle also applies to other species.

Behavior patterns that are adaptive in the evolutionary biologist's sense may be reinforcing in the psychologist's terms, as well. Perhaps natural selection has also favored the mental experiences that accompany adaptive behavior. It thus becomes almost a truism, once one reflects upon the question, that conscious awareness could have great adaptive value in the sense that this term is used by evolutionary biologists. The better an animal understands its physical, biological, and social environment, the better it can adjust its behavior to accomplish whatever goals may be important in its life, including those that contribute to its evolutionary fitness. The basic assumption of contemporary behavioral ecology and sociobiology, as the latter term is used by Wilson (1975) and many others, is that behavior is acted upon by natural selection along with morphological and physiological attributes. From this plausible assumption it follows that—insofar as any mental experiences animals have are significantly interrelated with their behavior—they, too, must feel the impact of natural selection. To the extent that they convey an adaptive advantage on animals, they will be reinforced by natural selection.

Arguments of this kind, which appeal to a presumed selective advantage, suffer from the limitation that a sufficiently fertile imagination can almost always find a plausible adaptive advantage for any observed trait. The very success of such arguments tends to undermine their strength, because if one can make up an equally plausible case for alternate explanations, there is little basis for preferring one explanation over any of the others. A stronger form of evolutionary argument is the converse position that any attribute with a selective *dis*advantage will almost certainly be eliminated unless it is genetically coupled with some compensating advantage. On this basis, we can at least argue that no selective drawbacks to conscious awareness have been demonstrated. Indeed, I am not aware that any have even been suggested.

In recent years, ethologists and ecologists have analyzed in great detail the behavioral strategies and tactics which animals employ in their daily affairs. These include how they distribute their time and energy in searching for food, in seeking mates, and even in managing their reproductive affairs in terms of the times selected for producing young, how many young to rear, and in general how to maximize the individual's contribution to the future gene pool of its species. As a result of these investigations, it is becoming clear that most animals behave in a relatively efficient manner, doing those things that tend to enhance their individual fitness, in the evolutionary biologist's sense of maximizing the proportion of future generations consisting of their offspring or collateral descendants (nephews, cousins, etc.). In such discussions, ethologists scarcely ever consider to what extent the animals exhibiting these effective tactics are consciously aware that they are doing so. Do the male lions or monkeys that displace previously dominant males from a group of females kill their predecessor's cubs with any understanding that they will thereby have a better chance of producing their own offspring at an earlier time or in greater numbers (Bertram, 1976; Hrdy, 1978)? Because similar sociobiological arguments can be applied to algae and to chimpanzees, the likelihood of awareness on the part of the actors cannot readily be judged from the nature and effectiveness of the tactics themselves, and other evidence, such as the sorts of complex and communicative behavior discussed in this book, must be relied on. But it may well be that animals which are consciously aware of their sociobiological goals can achieve them more effectively than would otherwise be the case.

The Nature and Nurture of Mental Experiences

One consequence of the view that awareness is simply one aspect of neurophysiological processes is to raise the nature-nurture question with regard to mental experiences themselves. To the extent that they are dealt with at all by scientists, it seems to be tacitly assumed that mental experiences result solely from individual experience and, in particular, from learning. This implication is clear in the statements of Pollio (1974), Maritain (1957), and Adler (1967), quoted in Chapters 5 and 6.

Whether or not we accept the behaviorists' axiom of psycho-neural identity (discussed from several viewpoints in the volume edited by Feyerabend and Maxwell, 1966), we should face up to the possibility that a nervous system might attain those properties leading to mental experiences primarily on the basis of genetic information. The development of mental experiences might depend on environmental influences only in the general and unspecific sense that DNA cannot lead to a complete animal without an environment that provides the necessary nourishment and other conditions. We might therefore conclude that the assumption of psychoneural identity leads to the likelihood of something akin to "innate ideas" in the philosophical sense.

The nature-nurture issue with respect to behavior has aroused some of the most violent passions and heated debates known among scientists. As is usual in such cases, balanced consideration strongly suggests that both sides are partly correct and that both individual experience and genetic heritage have significant effects on behavior (reviewed by J. L. Brown, 1975, and Marler et al., 1980). It is also obvious that the relative importance of these two major factors varies widely among behavior patterns and groups of animals. This means that the two extreme, all-or-nothing positions are clearly and equally untenable. Furthermore, interactions between genetic and environmental factors are of considerable importance, and this, together with the enormous difficulty of controlled experiments, makes it almost impossible to estimate their relative importance with anything approaching adequate accuracy. This does not mean that either can safely be assumed *a priori* to be all-important (or unimportant) in the absence of direct evidence of a sort that is rarely available at present.

Applying the same balanced approach to mental experience leads to a cautiously open mind concerning the possibility that both genetic and environmental influences, and interactions between them, may be important in the causation of mental processes, including conscious awareness. Because we know so little about mental experiences in other species, we can scarcely begin to attack the nature-nurture question, despite its potential importance. But we should not overlook the reality of the question, any more than it seems sensible to ignore the possible existence of mental expressions in more than one species.

10

A Possible Window
on the Minds of Animals

The aim of this chapter is to outline potential experiments that may offer some realistic hope of escaping from the difficulties discussed in previous chapters, and of beginning the exploration of scientific territory so unknown that its very existence has been seriously questioned. Scientists interested in behavior have tended in recent decades to call themselves behavioral scientists, a term that obviously is descriptive at two levels. They study what animals *do*, not the structure and function of their component organs, cells, or molecules; at the same time, they try to explain behavior, whenever possible, in terms of physiology or biochemistry. At a second and more subtle level, however, the term "behavioral," at least in scientific circles, strongly implies that these scientists are *not* concerned with thoughts or mental experiences of people or animals. Indeed, they often take a distinct pride of parsimony in sticking to observable behavior and ignoring mental phenomena. Since even thinking about mental experiences in animals has been largely tabooed, it is not surprising that ethologists and comparative psychologists have, as yet, learned very little about animal awareness.

A helpful point of departure is to compare the approach of an ethologist studying the communication behavior of another species with that of a hypothetical anthropologist making initial contact with a group of people whose language is totally unknown to him (Nance, 1975*). Because they are men, the anthropologist assumes that their sounds are indeed a form of speech. He notes correlations between their behavior patterns and their vocalizations and gestures, and he is certain to rely heavily on gestures in his own first attempts to communicate. The importance of nonverbal human communication is receiving increased recognition and is being effectively investigated from many viewpoints (Hinde, 1972; Krames et al., 1974; Benthal and Polhemus, 1975; Rosenthal et al., 1979). Our hypothetical anthropologist might take advantage of his knowledge of this field to encourage the people with whom he is trying to communicate to make at least some effort in the same direction, perhaps by pointing to persons or objects while uttering their names. Curiously enough, linguists and anthropologists have paid very little attention to the first steps needed to establish linguistic contact between people speaking languages unknown to one another (Hewes, 1974, 1975*).

It is conceivable that a diligent anthropologist might learn a great deal about a language by one-way visual and auditory observation of people as they speak. In principle, the same thing might be done by watching extensive television or motion-picture sequences provided with an adequate sound track. But even if success should be claimed from such a laborious effort, we would wish to test the claim by asking that direct, two-way communication be demonstrated. Indeed, we would have rather little confidence that the anthropologist had really learned the language until his knowledge passed this crucial test.

Studies of animal communication have so far remained almost entirely at a comparable level of correlated observations, except for some of the recent studies of signing apes reviewed in Chapter 4. We see that animals emit certain signals and observe how conspecifics respond. The signals may be sounds, visual patterns, scents, tactile vibrations, or even electric currents, and each channel presents different problems of monitoring what are suspected to be signals, or playing back artificial signals to observe any responses they may elicit. But except for teaching wordlike signs to apes, scarcely any effort has been made to move ahead to the next stage of *participatory* investigation. The Gardners and their successors succeeded in establishing far more complex two-way communication with apes than had previously seemed possible. One element in their success was the use of a communication channel (manual gestures) that chimpanzees learned to imitate far more readily than vocal sounds. They also had the advantage of being similar enough to their subjects morphologically so that acceptance as partners in social communication behavior was easier than between, say, man and dog. Yet millions of pet lovers have achieved limited forms of communication with dogs, cats, and other animals. An especially pertinent example is the rapport sometimes established between blind people and their guide dogs.

At first glance, the complexity and versatility of such communication seems to be more limited than the common vocabulary established between the Gardners and Washoe or between other investigators and other chimpanzees. But that impression may be misleading, because the signs which trained apes have learned to use more or less like words are known to us from the reports of the experimenters who worked long and hard to teach them to their subjects. On the other hand, nonverbal communication between people, animals, or between animals and men, is often so difficult to decode that we may easily underestimate the versatility of the system or the size of the equivalent vocabulary.

Participatory Investigation of Animal Communication

I should like to suggest that it is now the time to extend these approaches to other species, using methods analogous to those of anthropologists seeking to establish communication with conspecifics who are assumed to speak some language, but one which shares no words with any tongue known to the anthropologist. It may be necessary, almost literally, to talk back and forth with a communicating animal in order to verify the full meaning of its signals. Most animals are sufficiently different from men that an investigator is unlikely to be an acceptable social partner. If so, suitable models are called for. Such models must be similar enough to the animals in question, and versatile enough in their signaling behavior, to act as transmitters of information via whatever communication system is natural to the animals under study. Initially, the investigator would act as the receiver, through appropriate observations, but at a later stage in the development of such experiments he could manipulate the model to attempt *two-way communication*. Experimental participatory communication might, if suitably developed, provide us with a "window" through which to learn what the animal is thinking about. To become convincing, the data gathered in this way should, of course, be validated by replication, independent verification, and all the pertinent controls customary in experimental science.

Many animals respond well enough to various types of models to offer realistic hopes of eventual success (reviewed by Tinbergen, 1951, and by Marler and Hamilton, 1966). For example, some invertebrate and vertebrate animals react to their mirror images. In most cases, this does not lead to any prolonged interaction, but the mere fact that mirror images elicit any communication behavior at all tells us that visual signals can be mimicked with at least limited success. The fact that, as Gallup (1977) points out, the mirror image usually seems to be treated as another animal, rather than a self-image, could perhaps be turned to advantage. Closed-circuit video systems offer great promise for experimental dialogues, because the visual image can be manipulated much more extensively than is possible with mirrors. Stout and his colleagues have demonstrated that some important elements of aggressive display between glaucous-winged gulls *(Larus glaucescens)* can be elicited by relatively simple models with or without playback of vocalizations used in normal aggressive encounters (Stout et al., 1969; Stout and Brass, 1969; Stout, 1975; Galusha and Stout, 1977; Hayward et al., 1977; Amlaner and Stout, 1978). But these experiments have not yet been carried past the initial stage of showing that certain sounds or postures elicited stronger responses than did others. The experiments of Chauvin-Muckensturm and Pepperberg, discussed in Chapter 4, suggest that birds may have greater capabilities for symbolic communication than have yet been recognized.

Fireflies respond to brief flashes from a simple, hand-held flashlight (Lloyd, 1977, 1979), provided the flashes have the characteristic temporal patterns of the species. But there are many subtleties, even in this seemingly simple form of communication (Lloyd, 1980), and it might be worthwhile to attempt to use experimental dialogues to study how rich the system actually is.

Sounds are much easier to monitor, record, and reproduce with readily available instruments than are visual or chemical signals. Playback experiments have amply demonstrated their value in analyzing a wide variety of accoustical communication systems, such as the territorial calls exchanged between neighboring males in many species of birds. Such birds can easily be induced to exchange territorial calls with a tape recorder, but relatively little effort has yet been devoted to attempts at a more detailed two-way communication, comparable to the gestural communication established between the Gardners and Washoe. An important challenge is to analyze the details and nuances of animal communication, to inquire whether messages more subtle than the gross assertions of territoriality, for example, are exchanged. It is even easier to establish simple "dialogues" with electric fishes, because crude models provided with appropriate electrodes can readily emit and monitor their electrical communication signals (Hopkins, 1974, 1977, 1980; Westby, 1974). All these examples indicate that experimental dialogues between communicating animal and investigator are possible. The question is whether this approach can be exploited effectively to learn more about the animal's communicative abilities than the investigator knows in advance. This possibility can be evaluated only by new and enterprising experiments.

The possibility of communicating with a conspecific or with a model or mirror image sometimes appears to be reinforcing; that is, it seems that some animals like to engage in this kind of activity (Stevenson, 1969). Birds can certainly recognize each other as individuals through details of song pattern (W. J. Smith, 1969, 1977; Falls, 1969, 1978; Falls and Krebs, 1975*; Brooks and Falls, 1975; Kroodsma, 1978; Green and Marler, 1979), so that the basic requirements for elaboration of more detailed communication are clearly present. Wilson (1975), as well as Smith (1977), finds only a relatively small number of distinguishable messages, but a closer scrutiny via attempts at two-way communication might disclose a finer grain in the process. Perhaps careful analysis would reveal that combinations of signals carry a significantly different meaning from the individual signals.

Beer (1975, 1976) has discovered just this kind of finer grain in the communication behavior of laughing gulls *(Larus atricilla)*, including both sounds and visual signals conveyed by postures and motions. The long-call, which had previously been interpreted as a single communication signal, turns out to convey different messages under various circumstances, a possible case of context-dependence of the kind discussed by W. J. Smith (1968, 1969*, 1977). The long-call also serves in some cases to identify an individual gull so that it is recognized by its mate, neighbors, and chicks. Beer was led by his discoveries of the remarkable complexity and versatility of gull communication to conclude that "the long-call of Laughing Gulls is a form of display by means of which a gull can emphatically identify itself and convey a number of alternate messages with regard to itself . . . a long-call might thus signify 'I am your parent—come and get fed'; or 'I am your mate—let me sit on the eggs'; or 'I am your prospective mate—come and stay close'; or 'I am the occupier of this area—get out'. . . . the long-call . . . is semantically and pragmatically open' " (Beer, 1975). Elsewhere, Beer (1976) writes: "the recognition of greater complexity has resulted in, and in turn caused, changes in preconceived views about animal communication, including the models in terms of which animal communication has been thought about . . . linguistic analogies have, to some extent, taken the places previously occupied by causal and statistical models."

Dialogues with a Model Bee

The dance-communication system of honeybees provides another significant example, in which both the promise and the difficulties can readily be appreciated. Premack (1975) has argued that the dances cannot be accepted as truly significant communication unless they can be used to ask and answer questions, as discussed in Chapter 3. Because men and bees are so enormously different in size and morphology, to say nothing of gesturing capabilities, we need to develop an effective model bee. An important step in this direction was taken some years ago by Esch (1964; reviewed by Frisch, 1967). Esch employed relatively crude model bees, which he moved over the surface of the honeycomb in close approximations to the pattern of the waggle dance. Other bees often followed the model and emitted "stop signals"—sounds or vibrations that cause a dancer to stop and regurgitate food from her stomach. The followers then take up the food. If the model did not stop, it was attacked. Gould (1975c) has perfected another important aspect of such a model— the provision of an artificial substitute for regurgitation of sugar solution from the honey stomach. His model has been developed to the point at which other bees have accepted sugar solution from its artificial proboscis, although it is not yet clear whether the chemical stimulation resulting from the artificial trophallaxis is at all normal.

Much remains to be done before we will have available a model so acceptable to the bees that they act in accordance with the information it transmits. A successful model bee may have to display not only the correct mechanical motions, but also the appropriate odors. The latter may not yet be well-enough known to permit adequate simulation, but this gap could be closed by sufficient investigation. The technical requirements for a truly adequate model honeybee are thus formidable, but by no means impossible. Sounds and mechanical vibrations are important, and these, too, may have to be simulated with more precision than any experimenter has yet achieved. Vibrations that can be monitored with a small microphone placed near the dancer correlate well with the vigor of the dance. While the microphone responds to fluctuations in air pressure, the distance between the dancer and other bees is so small, compared to the wavelength of the apparent sound, that the signals are near-field effects, rather than typical propagated sound waves. Dances not accompanied by such sounds or vibrations are ineffective in recruiting other bees to fly out to the location signaled by the dancer. A further step would be for the experimenter to use the model as a vehicle for two-way communication with one or more bees. In such investigations, an experimenter could observe and interpret the dances directed at the model, and vary the model's behavior to generate whatever messages or responses were necessary to sustain an informative dialogue.

So far, this discussion has been limited to conservative descriptive statements, comparable to the pre-1940 observation that, under certain circumstances, many bees returned to a hive and carried out complex and variable motions which seemed to signify only some sort of general state of excitement. Having the benefit of Frisch's insights, we can now see a complex but orderly communication process at work. Let us entertain for the moment the hypothesis that the dancing conveys additional information, perhaps about the nature of the food source, along dimensions other than a single linear scale of desirability. If there is a more extensive and finer-grained two-way communication going on, how likely would we be to discover it by the type of observation customary to date? If, on the other hand, a model bee could be successful enough to be accepted by the real bees as a partner in communication behavior, we might learn about the existence and nature of more subtle elements in the communication system by participatory experiments analogous to the extensive and complex social interactions between the Gardners and Washoe.

This whole approach to the study of communication behav-

ior in animals is so poorly developed, even at the level of preliminary hypotheses, that it is difficult to anticipate where it might
lead. Perhaps the outcome of numerous and laborious experiments
would be a negative result, in that no additional kinds or degrees
of communication would be discovered. For example, Frisch
(1967) examined the possibility that bees might convey information about vertical, as well as horizontal, directions, but obtained a convincing negative answer. Even such negative results
would be of value in setting more precisely defined limits to the
communicative capabilities of the species in question. Alternately, however, it might be discovered that previously unsuspected messages are exchanged, and the scientific interest of
such discoveries is self-evident. If we consider the recent history
of this field, it is clear that far more complex communication
behavior has been found than any scientist would have ventured
to predict 30 years ago. Have we any reason to believe that
progress in this exciting area has reached a sudden end? We do
not know whether the properties of various animal communication systems, as they are now understood, are limited by the
capabilities of the communication systems themselves or by the
methods of investigation that have been employed to date.

One challenging approach would be direct "impersonation"
of a similar species, such as a chimpanzee, by an adequately
disguised experimenter using the gestures and sounds characteristic of chimpanzee communication. The disguise might have
to include not only visual appearance, but also chimpanzee
sounds and appropriate pheromonal perfumes. Jane Goodall
(1971) approached a state of acceptance by wild chimpanzees
without any attempt at morphological disguise, and she exchanged a few simple communication gestures, as others have
done with captive apes. A new generation of ambitiously pioneering ethologists might open up an enormously powerful new
science of participatory research in interspecies communication. First, however, they will have to overcome the feeling of
embarrassed outrage at this notion, and then laboriously develop the necessary techniques of disguise, imitation, and com-

municatory interaction.

Toward a Comparative Linguistics

Animal surrogates have been invaluable in the analysis and explanation of many biological phenomena, including some aspects of behavior, such as learning. The resulting knowledge and understanding has had many important applications to human medicine in particular and to human affairs in general. Biologists have often found that particular phenomena are more easily or more effectively investigated in certain especially suitable species, a principle often attributed to the physiologist August Krogh (Krebs, 1975). When a biological process is known only in our own species, its investigation is often difficult or impracticable. But when one or more animal surrogates are discovered, many kinds of experiments become feasible and scientific understanding is more readily attainable. It is common knowledge that cures or prevention of many human diseases have been made possible by this general procedure, and human welfare has been served in many other areas. The study of nutrition is a prime example. It is also obvious that this basic approach depends heavily on evolutionary continuity and the resulting confidence that the same basic principles can be applied to animal and human physiology.

Social communication behavior, broadly defined, is clearly of the highest significance in human affairs, comparable in importance to nutrition and physical health. But, unfortunately, we do not understand it nearly so well as we do many other areas of biology. This, in turn, suggests that the use of animal surrogates in experimental analyses of social-communication behavior could contribute significantly to a better understanding of human psychology, sociology, and even such apparently nonbiological disciplines as economics and philosophy. Such comparative analyses can never do the whole job, and any resulting conclusions must be checked against data obtained by studying human beings—just as new drugs or biochemical processes de-

veloped by animal experiments need careful checking with human subjects before their general application is wise. But many
advances in the biomedical sciences would have been greatly
impeded, if not hopelessly crippled, without the use of animal
surrogates for basic research. The pertinent analogy is to some
nutritional problem which people were so convinced was a
purely human phenomenon that they refused to test potential
dietary supplements on experimental animals.

In communication behavior, including human language, our
current situation is needlessly hampered by a tendency to deny
on theoretical grounds that evolutionary continuity exists and
that animal surrogates are thinkable. If one believes with Chomsky (1966) that human language is "based on an entirely different
principle," a whole avenue of investigation is blocked off. This is
surely an inefficient approach to an important and challenging
cluster of scientific problems. On the other hand, to the extent
that evolutionary continuity is significant in communication behavior, the entire momentum of comparative and experimental
science can be brought to bear on what has previously appeared
to be a uniquely specialized and almost unapproachable phenomenon.

Participatory two-way experiments of the general sort discussed above hold the potential of revealing important properties of animal communication systems that could be explored
only slowly, and with many uncertainties, by correlating signals
the animals exchange with behavior that is observed either si-
multaneously or subsequently. The immediate advantages lie in
the possibility of controlling the messages experimentally in
order to determine their effective content. This line of inquiry
should, in due course, enable ethologists to work out just what
messages are, in fact, exchanged between animals of a given
species under various conditions. But, in addition to this near
objective, there is also a more important, though distant, hope
that such methods will improve our ability to detect and analyze
whatever cognitive processes occur in the brains or minds of
animals. This means that a most important step toward under-

standing communication behavior will come when an investigator can ask questions and receive answers about any possible mental experiences (or, if behaviorists prefer, about covert communication behavior) in a given animal. Human language, despite all its limitations, does convey some information about subjective experiences in our fellow men. We need only extend to animals, with suitable modifications, the basic process by which we assess the mental experiences of our own species.

Human thoughts and words are closely linked, at the very least; and many philosophers have argued that they are essentially identical. Human mental experiences are often assumed to be so closely bound up with our species-specific neurophysiological mechanisms that we are not capable of understanding any mental, as distinct from neurophysiological, processes in other animals, even if such exist (Nagel, 1974). According to this view, should other species have feelings, hopes, beliefs, plans, or concepts of any sort—even very simple ones—they would take a form so different from our own that we could not recognize them. But basic matters like food-gathering, selecting a mate, caring for young, or avoiding predators are scarcely the monopolies of any one species. Therefore, thinking about these vital subjects may well be widespread among any species capable of flexibly adaptive behavior. In a similar vein, one might argue that animal thinking is linked equally closely to their communication systems.

A version of the widely held belief in human mental uniqueness might take the form of an analogy with computer hardware and software or programing (a suggestion discussed by Longuet-Higgins in Kenny et al., 1972, p. 25). It might be argued that central nervous systems, with their physiological mechanisms for information processing, storage, and retrieval, are analogous to the hardware of computers, and that human mental experiences are analogous to some particularly versatile type of software that might be considered unique to our own species. Insofar as this analogy is valid, only a small fraction of the programs are accessible to conscious experience. Presumably, this special

type of software would be closely linked to the powerful attributes of combinatorially productive human language. The argument might then go on to claim that other species have utterly
different software, which lacks the special features of human
mental programing. There are serious difficulties with this computer analogy, as lucidly explained by Searle (1980). But this
belief pattern resonates with contemporary enthusiasm for computers; and many of us tend to feel less demeaned by comparison
with computer systems than with "lower" animals.

As discussed by Premack (1975) and by Marler (1978), much
animal communication conveys an emotional or affective state,
such as fear. It is not surprising that the most easily recognized
types of animal communication concern matters of great importance to the animals themselves. Furthermore, ethologists are
more likely to detect and interpret communication which involves conspicuous departures from routine behavior, and this
is most likely to occur when the communicating animal shows a
high level of emotional arousal. Except for the dances of honeybees, some of the recruiting gestures of weaver ants, alarm calls
of vervet monkeys, and the signing recently taught to captive
chimpanzees, it is difficult to judge from available data whether
animal communication behavior also includes specific information about the nature of the object or situation responsible for
the emotional state conveyed. But rather than assuming the
absence of such information, as we now tend to do, it seems
advisable to consider this an open question to be investigated.
As pointed out in Chapter 3, communication with the property
of displacement provides important evidence.

A major objection to the return to the Darwinian viewpoint
that animals probably have mental experiences is the claim that
postulating such mental experiences does not immediately suggest definitive experiments which can decide among alternative
hypotheses. Another version of this criticism is that no falsifiable
hypotheses have been presented, that is, hypotheses which are
susceptible to experimental tests leading to a firm verdict of true
or false. By this is meant that alternative and almost, if not quite,

equally plausible interpretations are reasonably drawn from whatever outcome the proposed observations or experiments might yield. Thus, I may interpret the phenomena of bee dances as evidence that workers intentionally and consciously communicate information, whereas others, for example Gould (1979), conclude that the bees are complex automata, that the dances are correlates of certain physiological states, but that no conscious intention need be assumed to exist merely because symbolic communication is taking place. How can we work our way out of this dilemma?

One avenue of partial escape is to expand a suggestion advanced in Chapter 7. Many areas of science have made substantial and significant progress by formulating relatively imprecise hypotheses, gathering evidence pertinent to them, and gradually building up a persuasive, if not meticulously rigorous, case for an important conclusion. Darwin operated in essentially this manner, along with a host of his followers. The resulting conviction that men are genetically descended from animals had as great an impact on human thinking as any other scientific discovery. Only the Copernican revolution is a close rival, but altering fundamental views about the heavens was not as basic a change in our outlook on the universe as the recognition that we ourselves are genetically related to animals. Established religions vigorously resisted both revolutions, but later many accepted the newly recognized facts and adjusted their beliefs accordingly. In the case of biological evolution, that process is still not complete, and the issues discussed in this book are significantly debatable as part of that accommodation. Our full acceptance of the evolutionary continuity of mental experience may require another, and still unfinished, phase of the Darwinian revolution.

These general considerations may help explain our reluctance to consider seriously the possibility that we may share with other species not only anatomical, physiological, and biochemical attributes, but mental experiences, as well. Very few, if any, behavioral scientists overtly and explicitly reject animal awareness on religious grounds. But we are not immune to subtle residues of our cultural heritage, and faith in human uniqueness and superiority has long and pervasive roots. Such roots may well extend far deeper than historically recognizable religious beliefs. If the sociobiologists are correct that our evolutionary heritage has important effects on our behavior, the same may be true of our deep-seated beliefs. And, as discussed briefly in Chapter 9, a confident faith in the superiority of one's own group, especially of one's genetic relatives, may well have been highly adaptive.

In other scientific disciplines, new concepts or entities are proposed to help explain observations or the results of experiments without insisting on such adamant requirements for precise verification as those demanded by strict behaviorists for hypotheses about mental experiences. Why are we so insistent that mental experiences must be delivered in neatly concrete form, like an album filled with postage stamps, with all important particulars totally visible and susceptible to repeated and consistent measurement of every detail, before we allow ourselves to consider the possibility that they have any significant reality? Behaviorists seem content only with what might be called true-false or, at best, "multiple choice" experiments.

There is no doubt, however, that any scientist would be more satisfied if the concepts with which he deals can be stated in terms of data that are as fully observable and verifiable as possible—preferably more so than either neutrinos or mental images are at present. That is why I have based this discussion on human mental experiences, which most of us (even the strict behaviorists, when forced into a corner) recognize as real and significant. Language or, in a broader sense, communication behavior, is virtually the only method we have for learning about the subjective mental experiences of our fellow men. Furthermore, the newly recognized versatility of animal communication behavior opens up a closely comparable window by which we can hope to learn something, at least, about whatever mental experiences animals may have. In other words, the possibility of animal introspection is more than a will-o'-the-wisp; it is a potential method that already has been employed to a very limited degree by students of languagelike behavior in chimpanzees, and one that is ready for development and exploitation with other species to roughly the degree that they employ flexible communication systems. Clearly, the more versatile the communication system—which probably, but not necessarily, means the more symbolic it may be—and the more it has the property of displacement, the greater the opportunities for cognitive ethologists to employ it as a source of information about whatever mental images, intentions, and awareness animals may have.

By concentrating on communicating animals in the hope of learning something about their mental experiences, we should not overlook the broader possibility that some animals may also have mental experiences about which they do *not* communicate. Thus, for the foreseeable future, we may be able to discern only the tip of a large and significant iceberg. But this is strongly preferable to intentionally holding before our eyes an "iceberg-rejecting filter."

The pessimistic view that we are inherently incapable of communicating with animals involves the same sort of patronizing discouragement that I recall so vividly with regard to sun- and star-compass orientation or the use of echolocation to capture flying insects. A more hopeful prospect is offered by the success of comparative physiologists, comparative psychologists, and ethologists who have been able to analyze processes and mechanisms of animal orientation and communication, the very existence of which were undreamed-of until quite recently. Of course, to recognize the promise of a new approach is only the barest beginning. But the road now seems open, and I doubt that it is a dead-end street.

This chapter has emphasized the potential value of animal communication in general, and two-way participatory experiments in particular, as a source of information about whatever mental experiences animals may have. Of course, this is not the only source of evidence about the internal workings of animal brains or minds. Many other categories of versatile behavior can also be interpreted plausibly in terms of conscious intentions on the animals' part. These include constructing and using simple tools, as discussed by Beck (1980); building shelters or otherwise improving the immediate environment; aiding injured companions; sharing food; or hunting cooperatively. Inventive solutions of newly arisen problems and rapid spread of newly discovered behavior patterns, as in the "potato-washing culture" of Japanese macaques (Kawai, 1965; Wilson, 1975; Bonner, 1980; Mundinger, 1980), provide highly suggestive evidence of conscious awareness. In laboratory contexts, discriminative learning and problem-solving also provide strong evidence that animals think about what they are doing (Dücker and Rensch, 1977; Mackintosh, 1974; Hulse et al., 1978). Many scientists who work with primates or dogs recognize the likelihood of at least some conscious awareness. Indeed, one common response to the ideas outlined in this book has been "Why, yes, of course." As behavioral scientists recognize more clearly and widely the probable existence of mental experiences, we can proceed far more effectively to study their nature, causation, and significance.

This brings us back to a question discussed in previous chapters: Should a basic and fundamental distinction be made between a uniquely human use of language, on the one hand, and animal problem-solving on the other? Many critics of the ascription to signing apes of something similar to human language maintain that these animals have simply been trained to do certain things in order to solve specific problems for desired rewards, such as food or social companionship. Those critics contrast our use of language with the signing of apes by emphasizing that human speech is used not only for solving problems and obtaining rewards, but also, and more importantly, for a level of flexible and creative thinking that would otherwise be impossible. The importance of grammar or rules governing the combinations in which words are used and understood is a major consideration in this argument. It must be recognized that many animals communicate with each other, especially those that live in complex, mutually interdependent, social groups. Learning to respond appropriately to changing circumstances is also recognized as an attribute of almost all animals with even moderately complex nervous systems. But we are still strongly reluctant to credit any nonhuman animal with what is felt to be a precious perogative of humankind, centered around the creative use of combinatorially productive language.

Speculative Hypotheses Concerning the Central Role of Communicative Behavior

From all these many considerations of data and opinions emerges an important possibility which deserves consideration, even though no available evidence is adequate to evaluate it. Beginning with the widely held view that our own language is intimately related to our processes of thinking, or at least to some of the more important of these, I suggest that when social animals go to considerable trouble to communicate to their conspecifics in complex and versatile ways, they are sometimes consciously aware of the information they are busily communicating. To be sure, alternative, positivistic (behavioristic) explanations are always possible. But, as ingeniously worked out by Skinner (1957) and Epstein et al. (1980), the same can be said for human linguistic communication. Yet we know that our verbal and nonverbal communication is often accompanied by conscious awareness, and therefore these behavioristic arguments lose most of their force.

This simple notion that active and energetic communicative efforts can be accepted as evidence of conscious awareness will not be attractive initially to many behavioral scientists. Communicative behavior will seem to have no special status and provide no more convincing evidence for mental experience than complex discriminations, learning sets, or insightful problem-solving. All these categories of behavior also provide suggestive evidence, but specifically organized communicative behavior may well have a special status in the following sense.

When animals exert themselves in efforts to communicate, they direct their attention specifically toward conspecifics and often exchange communicative signals over considerable periods of time with mutually beneficial results.

These exchanges involve complex and adaptive interactions. Often there are long strings of back-and-forth exchanges in which a given animal's signals depend not only on its own state, but on the signals received from other animals. These communicative exchanges commonly require flexible versatility, calling on a wide variety of motor acts and remembered features of the surrounding world, as when individuals are recognized. Insofar as honeybees, weaver ants, or any other animals communicate about objects and events that are remote in space and time, they seem to be using more creative behavior than in most examples of discriminative learning and problem-solving. The very fact that conspecifics (often close genetic relatives) are involved in mutually advantageous exchanges of information adds a significant dimension to even the most complex behavior in which an individual animal may engage for its own immediate benefit.

These are not rigorous arguments. They are offered as intuitively appealing hypotheses, not fixed or dogmatic assertions immune from challenge or, I hope, from experimental testing. If we take for granted that our own mental experiences are real and significant, it seems more likely than not that because the central nervous systems of other animals are basically similar, they will share with our brains the capability of making possible at least some kinds of mental experiences. To conclude that nothing of the kind ever happens requires that we postulate an unparsimonious qualitative distinction between human brains and all those others that seem to have such similar structural and functional properties.

11

Summary and Conclusions

The communication behavior of certain animals is complex, versatile, and, to a limited degree, symbolic. The best-analyzed examples are the dances of honeybees and the signing of captive chimpanzees. These and other animal communication systems share many of the basic properties of human language, although in very much simpler form.

Language has generally been regarded as a unique attribute of human beings, different in kind from animal communication. But on close examination of this view, as it has been expressed by linguists, psychologists, and philosophers, it becomes evident that one of the major criteria on which this distinction has been based is the assumption that animals lack any conscious intent to communicate, whereas men know what they are doing. The available evidence concerning communication behavior in animals suggests that there may be no qualitative dichotomy, but rather a large quantitative difference in complexity of signals and range of intentions that separates animal communication from human language.

Human thinking has generally been held to be closely linked to language, and some philosophers have argued that the two are inseparable or even identical. To the extent that this assertion is accepted, and insofar as animal communication shares basic properties of human language, the employment of versatile communication systems by animals becomes evidence that they have mental experiences and communicate with conscious intent. The contrary view is supported only by negative evidence, which justifies, at the most, an agnostic position.

According to the strict behaviorists, it is more parsimonious to explain animal behavior without postulating that animals have any mental experiences. But mental experiences are also held by behaviorists to be identical with neurophysiological processes. Neurophysiologists have so far discovered no fundamental differences between the structure or function of neurons and synapses in man and other animals. Hence, unless one denies the reality of human mental experiences, it is actually parsimonious to assume that mental experiences are as similar from species to species as are the neurophysiological processes with which they are held to be identical. This, in turn, implies qualitative evolutionary continuity (though not identity) of mental experiences among multicellular animals.

The possibility that animals have mental experiences is often dismissed as anthropomorphic because it is held to imply that other species have the same mental experiences a man might have under comparable circumstances. But this widespread view itself contains the questionable assumption that human mental experiences are the only kind that can conceivably exist. This belief that mental experiences are a unique attribute of a single species is not only unparsimonious; it is conceited. It seems more likely than not that mental experiences, like many other characters, are widespread, at least among multicellular animals, but differ greatly in nature and complexity.

Awareness probably confers a significant adaptive advantage by enabling animals to react appropriately to physical, biological, and social events and signals from the surrounding world with which their behavior interacts.

Opening our eyes to the theoretical possibility that animals have significant mental experiences is only a first step toward the more difficult procedure of investigating their actual nature and importance to the animals concerned. Great caution is necessary until adequate methods have been developed to gather independently verifiable data about the properties and significance of any mental experiences animals may prove to have.

It has long been argued that human mental experiences can only be detected and analyzed through the use of language and introspective reports, and that this avenue is totally lacking in other species. Recent discoveries about the versatility of some animal communication systems suggest that this radical dichotomy may also be unsound. It seems possible, at least in principle, to detect and examine any mental experiences or conscious intentions that animals may have through the experimental use of the animal's capabilities for communication. Such communication channels might be learned, as in recent studies of captive apes, or it might be possible, through the use of models or by other methods, to take advantage of communication behavior which animals already use.

Recognizing the hazards of both positive and negative dogmatism in our present state of ignorance, how can ethologists handle the unsettled (and to some, unsettling) questions of animal awareness and consciousness? Open-minded agnosticism is clearly a necessary first step. Then, when the behavior of an animal suggests awareness, conscious intention, or simple forms of knowledge and belief, a second step might be to entertain the hypothesis that the particular animal under the given conditions may be aware of a certain fact or relationship or may be experiencing some feeling or perception. Granting that such hypotheses are difficult to test by currently available procedures, the tentative consideration of their plausibility might pave the way for thoughtful ethologists to devise improved methods to study when and where animal consciousness may occur and what its content may be. The future extension and refinement of two-way communication between ethologists and the animals they study offer the prospect of developing in due course a truly experimental science of cognitive ethology.

Bibliography

Adams, D. K. 1928. The inference of mind. *Psychol. Rev.* 35:235–252.

Adler, M. J. 1967. *The Difference of Man and the Difference it Makes*. New York: Holt, Rinehart & Winston.

Alston, W. P. 1972. Can psychology do without private data? *Behaviorism* 1:71–102. The conclusion is that it can.

Amlaner, C. J., and Stout, J. F. 1978. Aggressive communication by *Larus glaucescens*. VI. Interactions of territory residents with a remotely controlled, locomotory model. *Behaviour* 66:223–251.

Anshen, R. N. 1957. Language as Idea. In: Anshen, R. N. (Ed.), *Language: An Enquiry into its Meaning and Function*. New York: Harper.

Apter, M. J. 1970. *The Computer Simulation of Behavior*. London: Hutchinson University Library. A thoughtful discussion of the relationship of computer capability to consciousness.

Armstrong, E. A. 1949. Diversionary display. *Ibis* 91:88–97 and 179–188.

Ayala, J., and Dobzhansky, T. (Eds.). 1974. *Studies in the Philosophy of Biology, Reduction and Related Problems*. Berkeley: Univ. of California Press.

Beck, B. B. 1980. *Animal Tool Behavior*. New York: Garland STPM Press.

Beer, C. G. 1973a. Behavioral Components in the Reproductive Biology of Birds. In: *Breeding Biology of Birds*. Washington, D.C.: National Academy of Sciences, pp. 323–365.

———. 1973b. A View of Birds. In: Pick, A. (Ed.), *Minnesota Symposium on Child Biology*. Vol. 7. Minneapolis: Univ. of Minnesota Press.

———. 1975. Multiple Functions and Gull Displays. In: Baerends, G., Beer, C. G., and Manning, A. (Eds.), *Essays on Function and Evolution in Behaviour: A Festschrift for Professor Niko Tinbergen*. Oxford: Clarendon, Chapter 2.

———. 1976. Some Complexities in the Communication Behavior in Gulls. *Proc. of Conference on the Origins and Evolution of Language*. *Ann. N.Y. Acad. Sci.* 280:413–432.

Beninger, R. J., Kendall, S. R., and Vanderwolf, C. H. 1974. The ability of rats to discriminate their own behaviors. *Can. J. Psychol.* 28:79–91.

Bennett, J. 1964. *Rationality, Essay towards an Analysis*. London: Routledge and Kegan Paul.

———. 1978. Some remarks about concepts. *Behav. Brain Sci.* 1:557–560.

Benthall, J., and Polhemus, T. (Eds.) 1975. *The Body as a Medium of Expression*. New York: Dutton.

Berg, P., et al. 1974. Potential biohazards of recombinant DNA molecules. *Science* 185:303.

Bertram, B. C. R. 1976. Kin Selection in Lions and Evolution. In: Bateson, P. P. G., and Hinde, R. A. (Eds.), *Growing Points in Ethology*. Cambridge: Cambridge Univ. Press.

Bertrand, M. 1969. *The Behavioral Repertoire of the Stumptail Macaque, A Descriptive and Comparative Study*. Basel: S. Karger (Bibliotheca Primatologica No. 11).

Black, M. 1968. *The Labyrinth of Language*. New York: Praeger.

Blakemore, R. P. 1975. Magnetotactic bacteria. *Science* 190:377–379.

Bloomfield, L. 1933. *Language*. New York: Holt (Reprinted 1961 by Holt, Rinehart & Winston).

Bonner, J. T. 1980. *The Evolution of Culture in Animals*. Princeton, N.J.: Princeton Univ. Press.

Boring, E. G. 1963. *The Physical Dimensions of Consciousness*. New York: Dover.

Bourne, G. H. (Ed.). 1977. *Progress in Ape Research*. New York: Academic.

Boyle, D. G. 1971. *Language and Thinking in Human Development*. London: Hutchinson, pp. 29, 37, 152–153.

Bradshaw, J. W. S., Baker, R., and Howse, P. E. 1975. Multicomponent alarm pheromones of the weaver ant. *Nature* 258:230–231.

Brewer, W. F. 1974. There Is No Convincing Evidence for Operant or Classical Conditioning in Adult Humans. In: Weimer, W. B., and Palermo, D. S. (Eds.), *Cognition and the Symbolic Processes*. Hillsdale, N.J.: Erlbaum.

Bronowski, J., and Bellugi, U. 1970. Language, name, and concept. *Science* 168:669–673. The authors accept the early data from Washoe as demonstrating naming, but conclude that she does not use sentences and that her signing differs from human language in lacking rule-guided sequences of words.

Brooks, R. J., and Falls, J. B. 1975. Individual recognition by song in white-throated sparrows. I. Discrimination of songs of neighbors and strangers. *Can. J. Zool.* 53:879–888.

Brown, H. 1976. *Brain and Behavior*. New York: Oxford Univ. Press.

Brown, J. L. 1975. *The Evolution of Behavior*. New York: Norton.

Brown, R. 1958. *Words and Things*. New York: Free Press of Glencoe.

Brown, R. G. B. 1962. The aggressive and distraction behaviour of the western sandpiper *Ereunetes mauri*. *Ibis* 104:1–12.

Brown, S. R. 1972. A Fundamental Incommensurability between Objectivity and Subjectivity. In: Brown, S. R., and Brenner, D. J. (Eds.), *Sciences, Psychology, and Communication. Essays Honoring Williams Stephenson*. New York: Teachers College Press.

Bullock, T. H. 1973. Seeing the world through a new sense: electroreception in fish. *Am. Sci.* 61:316–325.

Busnel, R.-G., and Fish, J. F. (Eds.). 1980. *Animal Sonar Systems*. New York: Plenum.

Cassirer, E. 1953. *The Philosophy of Symbolic Forms*. Vol. 1. *Language*. New Haven: Yale Univ. Press, pp. 189 ff. "Even among the lower animals we encounter a great number of original sounds expressing feeling and sensation . . . cries of fear or warning, lures or mating calls. But between these cries and the sounds of designation and signification characteristic of human speech there remains a gap, a 'hiatus' which has been newly confirmed by sharper methods of observation of modern animal psychology." Here Cassirer cites Kohler (1925) in support of his conclusion that "The step to human speech, as Aristotle stressed, has been taken only when the purely significatory sound has gained primacy over sounds of affectivity and stimulation. . . ."

Cerella, J. 1979. Visual classes and natural categories in the pigeon. *J. Exp. Psychol. Hum. Percept. Perform.* 5:68–77.

Chauvin, R., and Chauvin-Muckensturm, B. 1980. *Behavioral Complexities.* New York: International Universities Press.

Chauvin-Muckensturm, B. 1974. Y a-t-il utilisation de signaux apprix comme moyen de communication chez le pic épeiche? *Rev. Comp. Animal* 9:185–207.

Cheney, D. L., and Seyfarth, R. M. 1980. Vocal recognition in free-ranging vervet monkeys. *Anim. Behav.* 28:362–367.

Chomsky, N. 1959. Review of Skinner's *Verbal Behavior. Language* 35:26–58. Reprinted with comments in Jacobovits, L. A., and Miron, M. S. (Eds.), *Readings in the Psychology of Language.* Englewood Cliffs, N.J.: Prentice-Hall.

———. 1966. *Cartesian Linguists.* New York: Harper & Row.

———. 1972. *Language and Mind.* New York: Harcourt Brace Jovanovich.

Cowgill, V. M. 1972. Death in *Perodicticus. Primates* 13:251–256.

Crane, J. 1975. *Fiddler Crabs of the World.* Princeton: Princeton Univ. Press.

Critchley, M. 1960. The Evolution of Man's Capacity for Language. In: Tax, S. (Ed.), *Evolution after Darwin.* Vol. II. Chicago: Univ. of Chicago Press.

Crook, J. H. 1980. *The Evolution of Human Consciousness.* New York: Oxford Univ. Press.

Croze, H. 1970. Searching image in carrion crows. *Z. f.Tierpsychol.* (Suppl. 5). Berlin: Parey.

Daanje, A. 1951. On the locomotory movements in birds and the intention movements derived from them. *Behaviour* 3:48–98.

Dennett, D. C. 1978. *Brainstorms, Philosophical Essays on Mind and Psychology.* Ann Arbor: Bradford Books.

Dethier, V. G. 1957. Communication by insects: The physiology of dancing. *Science* 125:331–336.

Dimond, S. J., and Blizard, D. A. (Eds.). 1977. Evolution and lateralization of the brain. *Ann. N.Y. Acad. Sci.* 299:1–501.

Dobzhansky, T. H. 1967. *The Biology of Ultimate Concern.* New York: New American Library.

Dücker, G., and Rensch, B. 1977. The solution of patterned string problems by birds. *Behaviour* 62:164–173.

Eccles, J. C. 1973. Brain, speech and consciousness. *Naturwissenschaften* 60:167–176.

———. 1974. Cerebral Activity and Consciousness. In: Ayala, F. J., and Dobzhansky, T. (Eds.), *Studies in the Philosophy of Biology, Reduction and Related Problems.* Berkeley: Univ. of California Press.

Edwards, P., and Pap, A. (Eds.). 1973. *A Modern Introduction to Philosophy*. 3rd ed. New York: Macmillan.

Elithorn, A., and Jones, D. (Eds.). 1973. *Artificial and Human Thinking*. San Francisco: Jossey-Bass, and Amsterdam: Elsevier.

Emlen, S. T. 1967. Migratory orientation in the Indigo Bunting, *Passerina cyanea*. *Auk* 84:309–342 and 463–489.

———. 1975. Migration: Orientation and Navigation. In: Farner, D. S., and King, J. R. (Eds.), *Avian Biology*. Vol. V. New York: Academic.

Epstein, R., Lanza, R. P., and Skinner, B. F. 1980. Symbolic communication between two pigeons *(Columba livia domestica)*. *Science* 207:543–545.

Esch, H. 1961. Über die Schallerzeugung beim Werbetanz der Honigbiene. *Z. vgl. Physiol.* 45:1–11.

———. 1964. Beiträge zum Problem der Entfernungsweisung in den Schwänzeltänzen der Honigbiene. *Z. vgl. Physiol.* 48:534–546.

Esch, H., and Bastian, J. A. 1970. How do newly recruited honey bees approach a food site? *Z. vgl. Physiol.* 68:175–181.

Estes, W. K. (Ed.). 1975. *Handbook of Learning and Cognitive Processes*. Vol. 1. *Introduction to Concepts and Issues*. Hillsdale, N.J.: Erlbaum.

Falls, J. B. 1969. Functions of Territorial Song in the White-Throated Sparrow. In: Hinde, R. A. (Ed.), *Bird Vocalizations*. London: Cambridge Univ. Press.

———. 1978. Bird Song and Territorial Behavior. In: Krames, L., Pliner, P., and Alloway, T. (Eds.), *Aggression, Dominance and Individual Spacing*. New York: Plenum.

Falls, J. B., and Krebs, J. R. 1975. Sequences of songs in repertoires of Western Meadowlarks *(Sturnella neglecta)*. *Can. J. Zool.* 53:1165–1178. Nonrandom responses to playback, conservatively analyzed.

Falls, J. B., and McNicholl, M. K. 1979. Neighbor-stranger discrimination by song in male blue grouse. *Can. J. Zool.* 57:457–462.

Feigl, H., Sellars, W., and Lehrer, K. 1972. *New Readings in Philosophical Analysis*. New York: Appleton-Century-Crofts. Here in a volume with scholarly, though divergent, views by several leading philosophers, Feigl maintains that the mind-body problem is not a pseudoproblem.

Fenton, M. B., and Fullard, J. H. 1979. The influence of moth hearing on bat echolocation strategies. *J. Comp. Physiol.* 132:77–86.

Feyerabend, P. K., and Maxwell, G. 1966. Mind, Matter, and Method. *Essays in the Philosophy of Science in Honor of Herbert Feigl*. Minneapolis: Univ. of Minnesota Press.

Fodor, J. A. 1968. *Psychological Explanation, An Introduction to the Philosophy of Psychology*. New York: Random House. A forceful attack on the behavioristic view of psychology, including: "Finally, it seems evident that behaviorism, considered as an account of theories of psychology, is simply a special case of operationalism considered as an account of scientific theories in general and that the former doctrine ought therefore to share in the discredit that has recently attached to the latter . . . the ultimate argument against behaviorism is simply that it seeks to prohibit *a priori* the employment of psychological explanations that may, in fact, be true."

――――. 1975. *The Language of Thought*. New York: Thomas Y. Crowell. Here Fodor argues further for the reality of mental processes.

Fodor, J. A., Bever, T. G., and Garrett, M. F. 1974. *The Psychology of Language*. New York: McGraw-Hill.

Fouts, R. S. 1973. Capacities for Language in the Great Apes. *Proc. IXth International Congress of Anthropological and Ethnological Sciences*. The Hague: Mouton & Co.

Fouts, R. S., and Rigby, R. L. 1977. Man-Chimpanzee Communication. In: Sebeok, T. A. (Ed.), *How Animals Communicate*. Bloomington: Indiana Univ. Press.

Fouts, R. S., Chown, W., and Goodwin, L. 1976. Translation from vocal English to American Sign Language in a chimpanzee (Pan). *Learn. Motiv*. 7:458–475.

Fraenkel, G. S., and Gunn, D. L. 1961. *The Orientation of Animals: Kineses, Taxes and Compass Reactions*. New York: Dover (reprint of original 1940 edition).

Frisch, K. von. 1923. Über die "Sprache" der Bienen. *Zool. Jahrb. Abt. allg. Zool. Physiol. Tiere*. 40:1–186.

――――. 1946. Die Tanze der Bienen. *Österr. Zool. Z*. 1:1–48.

――――. 1950. Die Sonne als Kompass im Leben der Bienen. *Experientia (Basel)* 6:210–221.

――――. 1967. *The Dance Language and Orientation of Bees*. (Translation by L. Chadwick.) Cambridge: Harvard Univ. Press.

――――. 1972. *Bees, their Vision, Chemical Senses and Language*. 2nd ed. Ithaca: Cornell Univ. Press.

――――. 1974. Decoding the language of the bee. *Science* 185:663–668.

Galaburda, A. M., Le May, M., Kemper, T. L., and Geschwind, N. 1978. Right-left asymmetries in the brain. *Science* 199:852–856.

Gallup, G. G., Jr. 1970. Chimpanzees: Self-recognition. *Science* 167:86–87.

――――. 1975. Towards an Operational Definition of Self-Awareness. In: Tuttle, R. (Ed.), *Socio-ecology and Psychology of Primates*. The Hague: Mouton.

――――. 1977. Self-recognition in primates. A comparative approach to the bidirectional properties of consciousness. *Am. Psychol*. 32:329–338.

Galusha, J. G., and Stout, J. F. 1977. Aggressive communication by *Larus glaucescens*. Part IV. Experiments on visual communication. *Behaviour* 62:222–235.

Gardner, B. T., and Gardner, R. A. 1969. Teaching sign language to a chimpanzee. *Science* 165:664–672.

――――. 1971. Two-way Communication with an Infant Chimpanzee. In: Schrier, A. M., and Stollnitz, F. (Eds.), *Behavior of Non-Human Primates*. Vol. IV. New York: Academic, Chapter 3.

――――. 1975. Evidence for sentence constituents in the early utterances of child and chimpanzee. *J. Exp. Psychol*. 104:244–267. Washoe used a vocabulary of at least 132 signs, including many that functioned as equivalents for nouns, proper and common, modifiers, markers such as "time to" verbs, and locatives.

―――. 1979. Two Comparative Psychologists Look at Language Acquisition. In: Nelson, K. E. (Ed.), *Children's Language*. Vol. 2. New York: Wiley.

Gazzaniga, M. S. 1975. Brain Mechanisms and Behavior. In: Gazzaniga, M. S., and Blakemore, C. (Eds.), *Handbook of Psychobiology*. New York: Academic, Chapter 19.

―――. 1979. *Handbook of Behavioral Neurobiology*. Vol. 2. *Neuropsychology*. New York: Plenum.

Globus, G. G., Maxwell, G., and Savodnik, I. (Eds.). 1976. *Consciousness and the Brain, a Scientific and Philosophical Inquiry*. New York: Plenum.

Glucksberg, S., and Danks, J. H. 1975. *Experimental Psycholinguistics; An Introduction*. Hillsdale, N.J.: Erlbaum. This book exemplifies the limitations of perspective which I feel should be reexamined and probably be relaxed. ". . . the dance of the honey bee shares none of the properties of human language systems."

Goldstein, K. 1957. The Nature of Language. In: Anshen, R. N. (Ed.), *Language: An Enquiry into its Meaning and Function*. New York: Harper, Chapter 2.

Goodall, J. van Lawick. 1968. Behavior of free-living chimpanzees of the Gombe Stream area. *Anim. Behav. Monogr.* 1:165–311.

―――. 1971. *In the Shadow of Man*. Boston: Houghton Mifflin.

―――. 1975. The Behavior of the Chimpanzee. In: Kurth, G., and Eibl-Eibesfeldt, I. (Eds.), *Hominisation und Verhalten*. Stuttgart: Gustav Fischer.

Goodfield, J. 1977. *Playing God, Genetic Engineering and the Manipulation of Life*. New York: Random House.

Gottlieb, G. 1971. *Development of Species Identification in Birds, an Inquiry into the Prenatal Determinants of Behavior*. Chicago: Univ. of Chicago Press.

Gould, J. L. 1974. Honey bee communication: Misdirection of recruits by foragers with covered ocelli. *Nature* 252:300–301.

―――. 1975a. Honey bee communication: The dance-language controversy. *Science* 189:685–693.

―――. 1975b. Communication of distance information by honey bees. *J. Comp. Physiol.* 104:161–173.

―――. 1975c. Honey Bee Communication: The Dance-Language Controversy. New York: The Rockefeller University, Ph.D. Thesis.

―――. 1976. The dance-language controversy. *Q. Rev. Biol.* 51:211–244.

―――. 1979. Do honeybees know what they are doing? *Nat. Hist.* 88:66–75.

―――. 1980. The case for magnetic orientation in birds and bees (such as it is). *Am. Sci.* 68:256–267.

Gould, J. L., Henery, J., and MacLeod, M. C. 1970. Communication of direction by the honey bee. *Science* 169:544–554.

Gould, J. L., Kirschvink, J. L., and Deffeyes, V. S. 1978. Bees have magnetic remanence. *Science* 201:1026–1028.

Gould, S. J. 1975. Man and other animals. *Nat. Hist.* 84(7):24–30.

Gramza, A. F. 1967. Responses of brooding nighthawks to a disturbance stimulus. *Auk* 84:72–86.

Green, S., and Marler, P. 1979. The Analysis of Animal Communication. In: Marler, P., and Vandenbergh, J. G. (Eds.), *Handbook of Behavioral Neurobiology*, Vol. 3. *Social Behavior and Communication*. New York: Plenum.

Gregg, L. W. 1974. *Knowledge and Cognition*. Potomac, Md.: Erlbaum.

Grice, H. P. 1957. Meaning. *Philos. Rev.* 66:377–388.

Griffin, D. R. 1946. Supersonic cries of bats. *Nature* 158:46–48.

———. 1950. Measurements of the ultrasonic cries of bats. *J. Acoust. Soc. Am.* 22:247–255.

———. 1952. Bird navigation. *Biol. Rev.* 27:359–393.

———. 1953. Bat sounds under natural conditions, with evidence for the echolocation of insect prey. *J. Exp. Zool.* 123:435–466.

———. 1958. *Listening in the Dark*. New Haven: Yale Univ. Press. (Reprinted 1974 by Dover, N.Y.)

———. 1973. Echolocation. In: Møller, A. R. (Ed.), *Basic Mechanisms in Hearing*. New York: Academic.

———. 1977. Expanding Horizons in Animal Communication. In: Sebeok, T. A. (Ed.), *How Animals Communicate*. Bloomington: Indiana Univ. Press.

———. 1978a. Humanistic Aspects of Ethology. In: Gregory, M. S., Silvers, A., and Sutch, D. (Eds.), *Sociobiology and Human Nature*. San Francisco: Jossey-Bass.

———. 1978b. Prospects for a cognitive ethology. *Behav. Brain Sci.* 1:527–538 (See also multiple commentaries and discussion, pp. 555–629 and 3:615–623, 1980).

———. 1978c. The Sensory Physiology of Animal Orientation. *Harvey Lectures*, Series 71. New York: Academic.

———. 1980. Early History of Research on Echolocation. In: Busnel, R.-G., and Fish, J. F. (Eds.), *Animal Sonar Systems*. New York: Plenum.

Griffin, D. R., and Suthers, R. A. 1970. Sensitivity of echolocation in cave swiftlets. *Biol. Bull.* 139:495–501.

Griffin, D. R., Webster, F. A., and Michael, C. 1960. The echolocation of flying insects by bats. *Anim. Behav.* 8:141–154.

Hailman, J. P. 1977a. *Optical Signals: Animal Communication and Light*. Bloomington: Indiana Univ. Press.

———. 1977b. Communication by Reflected Light. In: Sebeok, T. A. (Ed.), *How Animals Communicate*. Bloomington: Indiana Univ. Press.

Hampshire, S. 1959. *Thought and Action*. London: Chatto and Windus.

Harlow, H. F. 1949. The formation of learning sets. *Psychol. Rev.* 56:51–65.

Harnad, S. R., Steklis, H. D., and Lancaster, J. (Eds.). 1976. Origins and evolution of language and speech. *Ann. N.Y. Acad. Sci.* 280:1–914.

Hartridge, H. 1920. The avoidance of objects by bats in their flight. *J. Physiol. (Lond.)* 54:54–57.

Hartshorne, C. 1973. *Born to Sing. An Interpretation and World Survey of*

Bird Song. Bloomington: Indiana Univ. Press.

Hattiangadi, J. N. 1973. Mind and the origin of language. *Philos. Forum* 14:81–98.

Hayes, C. 1951. *The Ape in Our House*. New York: Harper.

Hayes, K., and Hayes, C. 1951. The intellectual development of a home-raised chimpanzee. *Proc. Am. Philos. Soc.* 95:105–109.

Haynes, P. F. R. 1976. Mental imagery. *Can. J. Philos.* 6:705–719.

Hayward, J. L., Gillett, W. H., and Stout, J. F. 1977. Aggressive communication by *Larus glaucescens*. Part V. *Orientation and sequence of behavior. Behaviour* 62:236–276.

Healy, A. F. 1971. Can chimpanzees learn a phonemic language? *J. Psycholinguist Res.* 2:167–170.

Hebb, D. O. 1974. What psychology is about. *Am. Psychol.* 29:71–79.

———. 1978. Behavioral evidence of thought and consciousness. *Behav. Brain Sci.* 1:577.

Heiligenberg, W. 1977. *Principles of Electrolocation and Jamming Avoidance in Electric Fish; A Neuroethological Approach*. New York: Springer.

Herman, L. M. (Ed.). 1980. *Cetacean Behavior Mechanisms and Functions*. New York: Wiley. An up-to-date and critical review of simple and complex behavior in many species of dolphins. Chapter 8 by Herman, "Cognitive Capacities of Dolphins," is especially pertinent.

Herrnstein, R. J., Loveland, D. H., and Cable, C. 1976. Natural concepts in pigeons. *J. Exp. Psychol. Anim. Behav. Processes* 2:285–302.

Hewes, G. W. 1974. Gesture language in culture contact. *Sign Language Studies* 1:1–34.

———. 1975. *Language Origins, A Bibliography*. 2nd ed. The Hague: Mouton. Vols. 1 and 2. A lengthy bibliography with many brief comments concerning the enormous literature on theories concerning the origin of human language.

Hinde, R. A. 1970. *Animal Behavior: A Synthesis of Ethology and Comparative Psychology*. New York: McGraw-Hill.

———. (Ed.). 1972. *Non-verbal Communication*. London: Cambridge Univ. Press.

Hockett, C. F. 1958. *A Course in Modern Linguistics*. New York: Macmillan.

Hockett, C. F., and Altmann, S. A. 1968. A Note on Design Features. In: Sebeok, T. A. (Ed.), *Animal Communication*. Bloomington: Indiana Univ. Press, Chapter 5.

Hölldobler, B. 1971. Recruitment behavior in *Camponotus socius* (Hym. Formicidae). *Z. vgl. Physiol.* 75:123–142.

———. 1974. Communication by tandem running in the ant *Camponotus sericeus. J. Comp. Physiol.* 90:105–127.

———. 1977. Communication in Social Hymenoptera. In: Sebeok, T. A. (Ed.), *How Animals Communicate*. Bloomington: Indiana Univ. Press.

———. 1978. Ethological aspects of chemical communication in ants. *Adv. Study Behav.* 8:75–115.

Hölldobler, B., Stanton, R. C., and Markl, H. 1978. Recruitment and food-retrieving behavior in *Novomessor* (Formicidae, Hymenoptera). *Behav. Ecol. Sociobiol.* 4:163–181.

Hölldobler, B., and Wilson, E. O. 1978. The multiple recruitment system of the African weaver ant *Oecophylla longinoda* (Latreille) (Hymenoptera, Formicidae). *Behav. Ecol. Sociobiol.* 3:19–60.

Holloway, R. L. 1974. Review of Jerison, H. J. *Evolution of the Brain and Intelligence*. *Science* 184:677–679.

Hopkins, C. D. 1974. Electrical communication in fish. *Am. Sci.* 62:426–437.

———. 1977. Electric Communication. In: Sebeok, T. A. (Ed.), *How Animals Communicate*. Bloomington: Indiana Univ. Press.

———. 1980. Evolution of electric communication channels of mormyrids. *Behav. Ecol. Sociobiol.* 7:1–13.

Hrdy, S. 1978. *The Langurs of Abu: Female and Male Strategies of Reproduction*. Cambridge: Harvard Univ. Press.

Hubbard, J. I. 1975. *The Biological Basis of Mental Activity*. Reading, Mass.: Addison-Wesley. A short and moderately elementary textbook with an innovative approach which leads the author to include a final chapter entitled "Brain and Mind." The conclusion reached is that no separate mental entities are necessary and that mental phenomena result directly from the activities of nervous systems.

Hulse, S. H., Fowler, H., and Honig, W. K. (Eds.). 1978. *Cognitive Aspects of Behavior in Animals*. Hillsdale, N.J.: Erlbaum.

Humphrey, N. 1977. Review of *The Question of Animal Awareness*. *Anim. Behav.* 25:521–522.

———. 1979. Nature's Psychologists. In: Josephson, B., and Ramachandra, B. S. (Eds.), *Consciousness and the Physical World*. New York: Pergamon. Reprinted in Sunderland, E., and Smith, M. T. 1980, *The Exercise of Intelligence*, New York: Garland STPM Press.

Hutchinson, G. E. 1976. Man talking or thinking. *Am. Sci.* 64:22–27.

Irwin, F. W. 1971. *Intentional Behavior and Motivation*. New York: Lippincott.

Jennings, H. S. 1906. *Behavior of Lower Organisms*. New York: Columbia Univ. Press. (Reprinted with preface by D. D. Jensen, Bloomington: Indiana Univ. Press, 1962.)

———. 1910. Diverse ideals and divergent conclusions in the study of behavior in lower organisms. *Am. J. Psychol.* 21:349–370.

———. 1933. *The Universe and Life*. New Haven: Yale Univ. Press.

Jolly, A. 1972. *The Evolution of Primate Behavior*. New York: Macmillan.

Kalmijn, A. J. 1971. The electric sense of sharks and rays. *J. Exp. Biol.* 55:371–383.

———. 1974. The Detection of Electric Fields from Inanimate and Animate Sources other than Electric Organs. In: Fessard, A. (Ed.), *Electroreceptors and other Specialized Receptors in Lower Vertebrates*. New York: Springer, Chapter 5.

———. 1978. Experimental Evidence of Geomagnetic Orientation in Elasmobranch Fishes. In: Schmidt-Koenig, K., and Keeton, W. T. (Eds.), *Animal Migration, Navigation and Homing*. New York: Springer.

Kawai, M. 1965. Newly acquired precultural behavior of the natural troop of Japanese monkeys on Kashima Islet. *Primates* 6:1–30.

Keeton, W. 1974. The orientational and navigational basis of homing in birds. *Adv. Study Behav.* 5:47–132.

Kellogg, W. N. 1961. *Porpoises and Sonar.* Chicago: Chicago Univ. Press.

Kenny, A. J. P., Longuet-Higgins, H. C., Lucas, J. R., and Waddington, C. H. 1972. *The Nature of Mind.* Edinburgh: Edinburgh Univ. Press.

———. 1973. *The Development of Mind.* Edinburgh: Edinburgh Univ. Press.

Kimble, G. A., and Perlmuter, L. C. 1970. The problem of volition. *Psychol. Rev.* 77:361–384. This paper begins with a pertinent quotation from C. S. Sherrington's *Integrative Action of the Nervous System* (New York: Scribner, 1906): "Yet it is clear, in higher animals especially so, that reflexes are under control . . . the reactions of reflex-arcs are controllable by mechanisms to whose activity consciousness is adjunct." Kimble and Perlmuter conclude ". . . that a voluntary act begins with an idea of the response to be performed. . . ." They also assume that "voluntary behavior is always learned," and they believe that "comparator mechanisms" must be present so that sensory input can be compared with the image assumed to be present internally.

Klein, D. B. 1970. *A History of Experimental Psychology.* New York: Basic Books.

Klima, E. S., and Bellugi, V. 1973. Teaching Apes to Communicate. In: Miller, G. A. (Ed.), *Communication, Language, and Meaning.* New York: Basic Books, Chapter 9.

Klopfer, P. H., and Hailman, J. P. 1967. *An Introduction to Animal Behavior.* Englewood Cliffs, N.J.: Prentice-Hall.

Knudsen, E. I., and Konishi, M. 1978. Space and frequency are represented separately in auditory midbrain of the owl. *J. Neurophysiol.* 41:870–884.

Köhler, W. 1918. Nachweis einfacher Strukturfunkitionen beim Schimpansen und beim Hauschuhn: Über eine neue Methode zur Untersuchung des bunten Farbensystems. *Abh. Preuss. Akad. Wiss.* 2:1–101. Translated under the title "Simple structural functions in the chimpanzee and in the chicken." In: Ellis, W. D., (Ed.), *A Source Book of Gestalt Psychology.* New York: Harcourt Brace, 1939, pp. 217–227.

———. 1925. *The Mentality of Apes.* Translated by E. Winter. London: K. Paul, Trench, Trubner; New York: Harcourt Brace.

Kosslyn, S. M. 1980. *Image and Mind.* Cambridge, Mass.: Harvard Univ. Press.

Kosslyn, S. M., Pinker, S., Smith, G. E., and Schwartz, S. P. 1979. On the demystification of mental imagery. *Behav. Brain Sci.* 2:535–581. This includes numerous comments and authors' replies.

Kramer, G. 1959. Recent experiments on bird orientation. *Ibis* 101:399–416.

Krames, L., Pliner, P., and Alloway, T. (Eds.). 1974. *Nonverbal Communication.* New York: Plenum.

Krebs, H. A. 1975. The August Krogh principle: "For many problems there is an animal on which it can be most conveniently studied." *J. Exp.*

Zool. 194:221–226.

Krebs, J. R. 1974. Colonial nesting and social feeding as strategies for exploiting food resources in the great blue heron *(Ardea herodias)*. *Behaviour* 51:99–134.

————. 1977. Review of *The Question of Animal Awareness*. *Nature* 266:792.

————. 1979. Foraging Strategies and their Social Significance. In: Marler, P. and Vandenbergh, J. G. (Eds.), *Handbook of Behavioral Neurobiology*. Vol. 3, *Social Behavior and Communication*. New York: Plenum.

Kreithen, M. L., and Keeton, W. T. 1974a. Detection of polarized light by the homing pigeon, *Columba livia*. *J. Comp. Physiol.* 89:83–92.

————. 1974b. Attempts to condition homing pigeons to magnetic stimuli. *J. Comp. Physiol.* 89:83–92.

Kroodsma, D. E. 1978. Aspects of Learning in the Ontogeny of Bird Song: Where, from whom, when, how many, which, and how accurately? In: Burghardt, G. M., and Bekoff, M. (Eds.), *The Development of Behavior: Comparative and Evolutionary Aspects*. New York: Garland STPM Press.

Langer, S. K. 1942. *Philosophy in a New Key*. New York: Pelican Books.

————. 1962. *Philosophical Sketches*. Baltimore: Johns Hopkins Press.

————. 1967. *Mind: An Essay on Human Feeling*. Vol. I. Baltimore: Johns Hopkins Press.

————. 1972. *Mind: An Essay on Human Feeling*. Vol. II. Baltimore: Johns Hopkins Press.

Lashley, K. S. 1923. The behavioristic interpretation of consciousness. *Psychol. Rev.* 30:237–272 and 329–353.

————. 1949. Persistent problems in the evolution of mind. *Q. Rev. Biol.* 24:28–42. (Reprinted in Beach, F. A., Hebb, D. O., Morgan, C. T., and Nissen, H. W. (Eds.). 1980. *The Neurophysiology of Lashley*. New York: McGraw-Hill.) Here Lashley restates the basic behavioristic position: "The questions of where mind or consciousness enters in the phylogenetic scale and of the nature of conscious experience as distinct from physiological processes are pseudo-problems, arising from misconceptions of the nature of the data revealed by introspection. A comparative study of the behavior of animals is a comparative study of mind, by any meaningful definition of the term."

————. 1951. The Problem of Serial Order in Behavior. In: Jeffress, L. A. (Ed.), *Cerebral Mechanisms in Behavior*. New York: Wiley. (Reprinted in Beach, F. A., Hebb, D. O., Morgan, C. T., and Nissen, H. W. 1960. *The Neuropsychology of Lashley*. New York: McGraw-Hill.)

————. 1958. Cerebral organization and behavior. *Proc. Assoc. Res. Nerv. Ment. Dis.* 36:1–18. (Reprinted in Beach, F. A., Hebb, D. O., Morgan, C. T., and Nissen, H. W. 1960, *The Neuropsychology of Lashley*. New York: McGraw-Hill).

Lawrence, D. H., and DeRivera, J. 1954. Evidence for relational transposition. *J. Comp. Physiol. Psychol.* 47:465–471.

Leger, D. W., and Owings, D. H. 1978. Responses to alarm calls by California ground squirrels: effects of call structure and material status. *Behav.*

Ecol. Sociobiol. 3:177–186.

Lehrman, D. 1953. A critique of Konrad Lorenz's theory of instinctive behaviour. *Q. Rev. Biol.* 28:337–363.

Lenneberg, E. H. 1971. Of language knowledge, apes, and brains. *J. Psycholinguist. Res.* 1;1–29.

———. 1975. Discussion of neuropsychological comparison between man, chimpanzee, and monkey. *Neuropsychologia* 13:125.

Levy, J. 1979. Strategies of Linguistic Processing in Human Split-Brain Patients. In: Fillmore, C. J., Kempler, D., and Wang, W. S.-Y. (Eds.), *Individual Differences in Language Ability and Language Behavior.* New York: Academic.

Lindauer, M. 1955. Schwarmbienen auf Wohnungssuch. *Z. vgl. Physiol.* 37;263–324. "Veilleicht ist dies mit ein entscheidender Punkt, um eine Einigung zustande zu bringen, dass die Spurbienen nicht hartnäckig bei ihrem ersten Urteil verbleiben, sondern nach kürzerer oder langerer Zeit verstummen und den weiteren Entscheid den Neulingen überlassen. 'Jetzt sollen die ihr Urteil abgehen.' . . . Das ist mit Abisicht anthropomorphistisch augsgedrukt. Selbstverständlich wird eine Spurbeine nicht bewusst ein Urteil bilden, wie es unserem menschlichen Begriff entspricht." Perhaps this is to be brought together with a crucial point, namely that the scout bees do not cling stubbornly to their first opinion, but, after a shorter or longer time, fall silent and yield further decision to the newcomers: "Now you give your opinion." This is intentionally expressed in anthropormophic terms. But it is self-evident that a scout bee does not consciously form an opinion corresponding to our human concepts.

———. 1971a. *Communication among Social Bees* (revised edition). Cambridge: Harvard Univ. Press.

———. 1971b. The functional significance of the honey bee waggle dance. *Am. Nat.* 105:89–96.

Lindauer, M., and Martin, H. 1972. Magnetic Effect on Dancing Bees. In: Galler, S. R., Schmidt-Koenig, K., Jacobs, G. J., and Belleville, R. E. (Eds.), *Animal Orientation and Navigation.* Washington, D.C.: U.S. Government Printing Office.

Linden, E. 1974. *Apes, Men, and Language.* New York: Dutton.

Lissmann, H. W. 1958. On the function and evolution of electric organs in fish. *J. Exp. Biol.* 35:156–191.

Lissmann, H. W., and Machin, K. E. 1958. The mechanism of object location in *Gymnarchus niloticus* and similar fish. *J. Exp. Biol.* 35:415–486.

Lloyd, J. E. 1975. Aggressive mimicry in *Photurus* fireflies: signal repertoires by femmes fatales. *Science* 187:452–453.

———. 1977. Bioluminescence and Communication. In: Sebeok, T. A. (Ed.), *How Animals Communicate.* Bloomington: Indiana Univ. Press.

———. 1979. Sexual Selection in Luminescent Beetles. In: Blum, M., and Blum, N. (Eds.), *Sexual Selection and Reproductive Competition.* New York: Academic.

———. 1980. Male *Photurus* fireflies mimic sexual signals of their females' prey. *Science* 210:669–671.

Loeb, J. 1900. *Comparative Physiology of the Brain and Comparative Psychology*. New York: Putnam.

———. 1912. *The Mechanistic Conception of Life*. Chicago: Univ. of Chicago Press. (Reprinted with preface by D. Fleming, Cambridge: Harvard Univ. Press, 1964.)

———. 1916. *The Organism as a Whole, from the Physicochemical Viewpoint*. New York: Putnam.

Lorenz, K. 1958. *Methods of Approach to the Problems of Behavior*. Harvey Lectures 1958–59. New York: Academic. (Reprinted in *Studies in Animal and Human Behavior*. Vol. II. Cambridge: Harvard Univ. Press, 1971.)

———. 1963. Haben Tiere ein subjectives Erleben? *Jahr. Techn. Hochs. Munchen*. (English translation reprinted in *Studies in Animal and Human Behavior*. Vol. II. Cambridge: Harvard Univ. Press, 1971.)

———. 1969. Innate Bases of Learning. In: Pribram, K. H. (Ed.), *On The Biology of Learning*. New York: Harcourt Brace and World. In a wideranging discussion of ethology and its implications, Lorenz reviews evidence that cultural transmission occurs in many examples of social behavior and communication among animals.

Mackintosh, N.J. 1974. *The Psychology of Animal Learning*. New York: Academic.

Malcolm, N. 1973. Thoughtless Brutes. *Proceedings and Addresses of American Philosophical Association*, 46: 5–20. Malcolm quotes Descartes as stating only that "it could not be *proved* either that animals do or that they do not have thoughts 'hidden in their bodies.'. . . But the idea that we cannot determine whether dogs have thoughts *in* them is a dreadful confusion. . . . The relevant question is whether they *express* thoughts. I think the answer is clearly in the negative. . . . The possession of language makes the whole difference. . . . An undertaking of trying to find out whether the dog did or didn't have that thought is not anything we understand.

"Descartes' notion was that speech is the only 'sign' of the presence of thought . . . that just conceivably animals may have thoughts of which they give no sign. This implies a looseness of connection between thought and the linguistic expression of thought, that is deeply disturbing. . . . The relationship between language and thought must be . . . so close that it is really senseless to conjecture that people may *not* have thoughts, and also really senseless to conjecture that animals *may* have thoughts."

Maritain, J. 1957. Language and the Theory of Sign. In: Anshen, R. N. (Ed.), *Language: An Inquiry into its Meaning and Function*. New York: Harper & Row, Chapter 5.

Markl, H. 1967. Die Verständigung durch Stridulationssignale bei Blattschneiderameisen. I. Die Biologische Bedeutung der Stridulation. *Z. vgl. Physiol*. 57:299–330.

———. 1968. Die Verständigung durch Stridulationssignale bei Blattschneiderameisen. II: Erzugung und Eigenschaften der Signale. *Z. vgl. Physiol.* 60:103–150.

————. 1970. Die Verständigung durch Stridulationssignale bei Blattschneiderameisen. III: Die Empfindlichkeit für Substratvibrationen. *Z. vgl. Physiol.* 69:6–37.

Markl, H., Lang, H., and Wiese, K. 1973. Die Genauigkeit der Ortung eines Wellenzentrums durch den Rückenschwimmer *Notonecta glauca* L. *J. Comp. Physiol.* 86:359–364.

Marler, P. 1968. Visual Systems. In: Sebeok, T. A. (Ed.), *Animal Communication.* Bloomington: Indiana Univ. Press, Chapter 7.

————. 1969. Animals and Man: Communication and its Development. In: Roslansky, J. D. (Ed.), *Communication.* Amsterdam: North-Holland, pp. 26–61.

————. 1974. Animal Communication. In Krames, L., Pliner, P., and Alloway, T. (Eds.), *Nonverbal Communication.* New York: Plenum, Chapter 2.

————. 1978. Perception and Innate Knowledge. In: Heidcamp, W. H. (Ed.), *The Nature of Life.* Baltimore: University Park Press.

Marler, P. R., Dooling, R. J., and Zoloth, S. 1980. Comparative Perspectives on Ethology and Behavioral Development. In: Bornstein, M. H. (Ed.), *Comparative Methods in Psychology.* Hillsdale, N.J.: Erlbaum.

Marler, P., and Hamilton, W. H. 1966. *Mechanisms of Animal Behaviour.* New York: Wiley.

Marler, P., and Tenaza, R. 1977. Signaling Behavior in Apes with Special Reference to Vocalization. In: Sebeok, T. A. (Ed.), *How Animals Communicate.* Bloomington: Indiana Univ. Press.

Marshall, A. J. 1954. *Bower-birds, Their Displays and Breeding Cycles.* London: Oxford Univ. Press.

Martin, H., and Lindauer, M. 1973. Orientierung im Erdmagnetfeld. *Fortschr. Zool.* 21:211–228.

Mason, W. A. 1976. Review of *The Question of Animal Awareness. Science* 194:930–931.

Matthews, G. V. T. 1968. Bird Navigation. 2nd ed. London: Cambridge Univ. Press.

McGrew, W. C. 1974. Tool use by wild chimpanzees in feeding upon driver ants. *J. Hum. Evol.* 3:501–508.

McGrew, W. C., Tutin, C. E. G., and Baldwin, P. J. 1979. Chimpanzees, tools and termites: cross-cultural comparisons of Senegal, Tanzania, and Rio Muni. *Man* 14:185–214.

McGuigan, F. J., and Schoonover, R. A. 1973. *The Psychophysiology of Thinking: Studies of Covert Processes.* New York: Academic.

McMullan, E. M. 1969. Man's Effort to Understand the Universe. In: Roslansky, J. D. (Ed.), *The Uniqueness of Man.* Amsterdam: North-Holland. McMullan takes it for granted that at least some mental experiences occur in animals, but he argues against extension to nonhuman animals of subjectivity of conscious awareness. He objects to what he considers two extreme views, one labeled "behaviorist" and the other "ethologist," which he defines by hypothetical quotations: Behaviorist: "there is no more to thinking than the overt behavior elicited and the accompanying neurological changes"; and Ethologist: "the honey-bee's use of language indicates a symbolic intelligence of the sort at work in human language, only at a simpler level." McMullan asserts that both extreme views "tend to smooth out the differences between organisms." He rejects the dances of bees as a language because "First, they are species-specific, inherited not learnt. Their use is instinctive, not reflective. Honey-bees of one species will not be able to 'follow' the language of another species, nor can they learn it."

Menzel, E. W., Jr. 1974. A Group of Young Chimpanzees in a One-Acre Field. In: Schrier, A. M., and Stollnitz, F. (Eds.), *Behavior of Nonhuman Primates*. Vol. 5. New York: Academic.

———. 1978. Cognitive Mapping in Chimpanzees. In: Hulse, S. H., Fowler, H., and Honig, W. K. (Eds.), *Cognitive Aspects of Behavior in Animals*. Hillsdale, N.J.: Erlbaum.

Menzel, E. W., Jr., and Halperin, S. 1975. Purposive behavior as a basis for objective communication between chimpanzees. *Science* 189:652–654.

Menzel, E. W., Jr., and Johnson, M. K. 1976. Communication and cognitive organization in humans and other animals. *Ann. N.Y. Acad. Sci.* 280:131–142.

Michener, C. D. 1974. *The Social Behavior of the Bees*. Cambridge: Harvard Univ. Press.

Miller, G. A. 1962. *Psychology, the Science of Mental Life*. New York: Harper & Row.

———. 1967. *The Psychology of Communication*. New York: Basic Books.

Miller, G. A., Galanter, E., and Pribram, K. H. 1960. *Plans and the Structure of Behavior*. New York: Holt, Rinehart & Winston, Chapter 11.

Mittelstaedt, H. 1972. Kybernetik der Schwereorientierung. *Verh. Dtsch. Zool. Ges.* (1971) 185–200.

Moehres, F. P., and Oettingen-Spielberg, T. 1949. Versuche über die Nahorientierung und Heimfindevermögen der Fledermause. *Verh. Dtsch. Zool. Ges.* 1949: 248–252.

Möglich, M., and Hölldobler, B. 1975. Communication and orientation during foraging and emigration in the ant *Formica fusca. J. Comp. Physiol.* 101:275–288.

Monod, J. L. 1975. On Molecular Theory of Evolution. In: Harre, R. (Ed.), *Problems of Scientific Revolution, Progress and Obstacles to Progress in the Sciences*. London: Oxford Univ. Press.

Morgan, C. L. 1894. *An Introduction to Comparative Psychology*. London: Scott.

Morris, C. 1946. *Signs, Language, and Behavior*. Englewood Cliffs, N.J.: Prentice-Hall. (Reprinted 1955 by Braziller). Morris uses as an example of the difference between his usage of sign and symbol a situation in which a dog has been conditioned to respond to the sound of a buzzer by going to some place other than the buzzer to obtain food. ". . . the advantage of such symbols is found in the fact that they occur in the

absence of signals produced by the environment; an action or state of
the interpreter itself becomes (or produces) a sign guiding behavior with
respect to the environment. So if a symbol operates in the dog's behavior, the symbol could take the place in the control of behavior which the
buzzer formerly exercised: hunger cramps for instance might themselves
come to be a sign (that is, a symbol) of food at the customary place." One
may well decide that such semantic exercises as these are of dubious
value in dealing with complex communication behavior such as the waggle dances of honeybees, which were unknown when Morris developed
his definitions. Conversely, if it seems more reasonable to consider the
waggle dance as a signal-process rather than a symbol-process in Morris's terms, this will have no important bearing on the principal considerations under discussion.

Mowrer, O. H. 1950. *Learning Theory and Personality Dynamics*. New
York: Ronald.

———. 1952. The autism theory of speech development and some clinical
application. *J. Speech Hear. Disord*. 17:263–268.

———. 1954. The psychologist looks at language. *Am. Psychol*. 9:660–694.

———. 1958. Hearing and speaking: An analysis of language learning. *J.
Speech Hear. Disord*. 23:143–152.

———. 1960a. *Learning Theory and Behavior*. New York: Wiley.

———. 1960b. *Learning Theory and Symbolic Processes*. New York: Wiley.

Mundinger, P. C. 1970. Vocal imitation and individual recognition of finch
calls. *Science* 168:480–482.

———. 1980. Animal cultures and a general theory of cultural evolution.
Ethol. Sociobiol. 1:183–223.

Nagel, T. 1974. What is it like to be a bat? *Philos. Rev*. 83:435–450.

Nance, J. 1975. *The Gentle Tasaday, A Stone Age People in the Philippine
Rain Forest*. New York: Harcourt Brace Jovanovich. The discussion
included in this charming book of first contacts with a newly discovered
tribe indicates that a few words were shared by the Tasaday and the
hunter who first made contact with them. The situation in which adult
human beings suddenly come into contact without sharing any words
whatsoever seems at the present time to be so rare that it has received
very little attention from anthropologists or linguists, except for Hewes
(1974, 1975).

Natsoulas, T. 1978a. Toward a model for consciousness in the light of B. F.
Skinner's contribution. *Behaviourism* 6:139–175.

———. 1978b. Residual subjectivity. *Am. Psychol*. 33:269–283.

Nebes, R. D., and Sperry, R. W. 1971. Hemispheric deconnection syn-
drome with cerebral birth injury in the dominant arm area. *Neuropsychologia* 9:247–259.

Neisser, U. 1967. *Cognitive Psychology*. New York: Appleton-Century-Crofts.

Neville, H. J. 1976. The Functional Significance of Cerebral Specialization.
In: Rieber, R. W. (Ed.), *The Neuropsychology of Language*. New York:
Plenum.

Nicholas, J. M. (Ed.). 1977. *Images, Perception, and Knowledge*. Dordrecht,

Holland: Reidel.

Norris, K. S. (Ed.). 1966. *Whales, Dolphins, and Porpoises*. Berkeley: Univ. of California Press.

Nottebohm, F. 1977. Neural Asymmetries in the Vocal Control of the Canary. In: Harnad, S. R., and Doty, R. W. (Eds.), *Lateralization in the Nervous System*. New York: Academic.

————. 1979. Origins and Mechanisms in the Establishment of Cerebral Dominance. In: Gazzaniga, M. S. (Ed.), *Handbook of Behavioral Neurobiology*. Vol. 2. *Neuropsychology*. New York: Plenum.

Nottebohm, F., and Nottebohm, M. 1976. Phonation in the orange-winged Amazon parrot, *Amazona amazonica*. *J. Comp. Physiol*. 108:157–170.

O'Keefe, J., and Nadel, L. 1978. *The Hippocampus as a Cognitive Map*. New York: Oxford Univ. Press.

Olton, D. S. 1979. Mazes, maps, and memory. *Am. Psychol*. 34:583–596.

Olton, D. S., and Samuelson, R. J. 1976. Remembrance of places passed: spatial memory in rats. *J. Exp. Psychol.: Anim. Behav. Processes* 2:97–116.

Owings, D. H., and Leger, D. W. 1980. Chatter vocalizations of California ground squirrels: predator- and social-role specificity. *Z. Tierpsychol*. 54:163–184.

Owings, D. H., and Virginia, R. A. 1978. Alarm calls of California ground squirrels. *Z. Tierpsychol*. 46:58–70.

Patterson, F. G. 1978. The gestures of a gorilla: language acquisition in another pongid. *Brain Lang*. 5:72–97.

Pepperberg, I. M. Functional vocalizations of an African Grey parrot *(Psittacus erithacus)*. *Z. Tierpsychol*. (in press).

Pfungst, O. 1911. *Clever Hans, the Horse of Von Osten*. (With preface of J. R. Angell and introduction by C. Stumpf.) New York: Holt, Rinehart & Winston. (Reprinted 1965 in English translation by C. L. Rahn and introduction by R. Rosenthal.)

Piaget, J. 1971. *Biology and Knowledge: An Essay on the Relations between Organic Regulations and Cognitive Processes*. Chicago: Univ. of Chicago Press. (Translation by B. Walsh of *Essai sur les Relations entre les Regulations Organiques et les Processus Cognitifs*. Paris: Gallimard, 1967.)

Pollio, H. R. 1974. *The Psychology of Symbolic Activity*. Reading, Mass.: Addison-Wesley. A review of honeybee communication which the author does not accept as a language.

Polten, E. P. 1973. *Critique of the Psycho-Physical Identity Theory*. The Hague: Mouton. This book presents counterarguments to those of Feigl. Animals are denied free will (page 163) even though the author admits on page 230 that "we cannot be sure of the exact type of internal consciousness of various animals." On page 160, he states: "It does seem indeed true that animals are not *aware* what they are experiencing or doing, cannot conceptualize or think." On page 129, Polten quotes Hegel: "The only mere physicists are the animals: they alone do not think: while man is a thinking being and a born metaphysician."

Popper, A. N. 1980. Sound emission and defection by delphinids. In: Herman, L. M. (Ed.), *Cetacean Behavior, Mechanisms and Functions*. New York: Wiley, Chapter 1.

Popper, K. R. 1972. *Objective Knowledge. An Evolutionary Approach*. London: Oxford Univ. Press. On page 74 Popper tells us "Animals, although capable of feelings, sensations, memory, and thus of consciousness, do not possess the full consciousness of self which is one of the results of human language. . . ." On page 122 he continues, "The most important functions of dimensions of the human language (which animal languages do not possess) are the descriptive and argumentative functions."

————. 1974. Scientific Reduction and the Essential Incompleteness of all Science. In: Ayala, F. J., and Dobzhansky, T. (Eds.), *Studies in the Philosophy of Biology, Reduction and Related Problems*. Berkeley: Univ. of California Press, Chapter 16. "Though this behaviorist philosophy is quite fashionable at present, a theory of the nonexistence of consciousness cannot be taken any more seriously, I suggest, than a theory of the nonexistence of matter."

Popper, K. R., and Eccles, J. C. 1977. *The Self and Its Brain*. New York: Springer.

Posner, M. I. 1978. *Chronometric Explorations of Mind*. Hillsdale, N.J.: Erlbaum.

Premack, D. 1975. Symbols Inside and Outside of Language. In: Kavanagh, J. F., and Cutting, J. E. (Eds.), *The Role of Speech in Language*. Cambridge: M.I.T. Press.

————. 1976. *Intelligence in Ape and Man*. Hillsdale, N.J.: Erlbaum.

————. 1978. Chimpanzee theory of mind: Part II. The evidence for symbols in chimpanzee. *Behav. Brain Sci.* 1:625–629.

————. 1980. Representational Capacity and Accessibility of Knowledge: The Case of the Chimpanzee. In: Piatelli-Palmarini, M. (Ed.), *Language and Learning, the Debate between Jean Piaget and Noam Chomsky*. Cambridge: Harvard Univ. Press, Chapter 9.

Premack, D., and Woodruff, G. 1978. Does the chimpanzee have a theory of mind? *Behav. Brain Sci.* 1:515–526. See also discussion pp. 555–629.

Pribram, K. H. 1971. *Languages of the Brain*. Englewood Cliffs, N.J.: Prentice-Hall.

Price, H. H. 1938. Our evidence for the existence of other minds. *Philosophy*. 13:425–456.

Pyles, T. 1971. *The Origins and Development of the English Language*. 2nd ed. New York: Harcourt Brace Jovanovich.

Razran, G. 1971. *Mind in Evolution, an East-West Synthesis of Learned Behavior and Cognition*. Boston: Houghton-Mifflin.

Rensch, B. 1971. *Biophilosophy*. (Translation of 1968 German edition by C. A. M. Sym.) New York: Columbia Univ. Press. Rensch accepts the likelihood that animals have mental experiences and consciousness, but feels that animal concepts differ from human concepts in not being combined with words.

Ristau, C. A., and Robbins, D. 1979. A threat to man's uniqueness? Language and communication in the chimpanzee. *J. Psycholinguist. Res.*

8:267–300.

———. 1981a. Language in the Great Apes: A Critical Review. In: Rosenblatt, J. S., Hinde, R. A., Beer, C., and Busnel, M-C (Eds.), *Advances in the Study of Behavior*. Vol. 12. New York: Academic.

———. 1981b. Cognitive Aspects of Ape Language Experiments. In: Griffin, D. R. (Ed.), *Animal Mind—Human Mind*. Berlin: Dahlem Konferenzen (in press).

Robinson, D. N. 1973. *The Enlightened Machine: An Analytical Introduction to Neurophysiology*. Encino, California: Dickenson. (Reprinted 1980 by Columbia Univ. Press.) In the final chapter of this semipopular survey of neurophysiology, the author begins to define language in terms of a parrot's "Polly wanna cracker," honeybee dances, and clouds that "inform us of impending rain, and when they take on a certain texture and mass, they tell us to find shelter. We reject the clouds' communication as language because, we insist, the clouds do not *know* what they are telling us; they do not *intend* to inform us. . . . There is one thing that babies, puppies, parrots, and clouds have in common: *They cannot lie*."

Roeder, K. 1970. Episodes in insect brains. *Am. Sci.* 58:378–389.

Roeder, K., and Treat, A. 1957. Ultrasonic reception by the tympanic organ of noctuid moths. *J. Exp. Zool.* 134:127–157.

Rohles, F. H., and Devine, J. V. 1966. Chimpanzee performance on a problem involving the concept of middleness. *Anim. Behav.* 14:159–162.

———. 1967. Further studies of the middleness concept with the chimpanzee. *Anim. Behav.* 15:107–112.

Rosenthal, R., Hall, J. A., DiMatteo, M. R., Rogers, P. L., and Archer, D. 1979. *Sensitivity to Nonverbal Communication, the PONS Test*. Baltimore, Md.: Johns Hopkins Univ. Press.

Rosin, R. 1980a. The honey bee "language" controversy. *J. Theoret. Biol.* 72:589–602.

———. 1980b. The honey-bee "dance language" hypothesis and the foundations of biology and behavior. *J. Theoret. Biol.* 87:457–481.

Rumbaugh, D. M. (Ed.). 1977. *Language Learning by a Chimpanzee*. New York: Academic.

Rüppel, G. 1969. Eine "Lüge" als gerichtete Mitteilung beim Eisfuchs *(Alopex lagopus L.)*. *Z. Tierpsychol.* 26:371–374.

Ryle, G. 1949. *The Concept of Mind*. London: Hutchinson.

Sarles, H. 1975. Language and Communication. II: The View from '74. In: Pliner, P., Krames, L., and Alloway, T. (Eds.), *Nonverbal Communication of Aggression*. New York: Plenum.

Sauer, E. G. F. 1957. Die Sternenorientierung nächtich ziehender Grasmücken *(Sylvia atricapilla, borin* und *curruca)*. *Z. Tierpsychol.* 14:29–70.

Savage-Rumbaugh, E. S., Rumbaugh, D. M., and Boysen, B. 1978. Linguistically mediated tool use and exchange by chimpanzees *(Pan troglodytes)*. *Behav. Brain Sci.* 4:539–554.

———. 1980. Do apes use language? *Am. Sci.* 68:49–61.

Savage-Rumbaugh, E. S., Wilkerson, B. J., and Bakeman, R. 1977. Spontaneous Gestural Communication among Conspecifics in the Pigmy Chimpanzee *(Pan paniscus)*. In: Bourne, G. H. (Ed.), *Progress in Ape Research*. New York: Academic.

Schaffer, J. A. 1975. Philosophy of mind. *Encyclopedia Brittanica Macropedia*, Vol. 12, pp. 224–233. Chicago: Encyclopedia Brittanica.

Schmidt-Koenig, K. 1965. Current problems in bird orientation. *Adv. Study Behav.* 1:217–278.

Schmidt-Koenig, K., and Keeton, W. T. (Eds.). 1978. *Animal Migration, Navigation, and Homing*. New York: Springer.

Schneirla, T. C. 1966. Behavioral development and comparative psychology. *Q. Rev. Biol.* 41:283–302.

Schultz, D. 1975. *A History of Modern Psychology*. 2nd ed. New York: Academic.

Scriven, M. 1963. The Mechanical Concept of Mind. In: Sayre, K. M., and Crosson, F. J. (Eds.), *The Modeling of Mind, Computers and Intelligence*. Notre Dame, Ind.: Notre Dame Press. "Consciousness is, it now seems to me, automatically guaranteed by the capacity to categorize and discuss one's own reactions, beliefs, etc." (p. 254).

Searle, J. R. 1980. Minds, brains and programs. *Behav. Brain Sci.* 3:417–457. Includes multiple commentaries and authors' replies.

Sebeok, T. A. (Ed.). 1977. *How Animals Communicate*. Bloomington: Indiana Univ. Press.

Sebeok, T. A., and Umiker-Sebeok, J. (Eds.). 1980. *Speaking of Apes, A Critical Anthology of Two-Way Communication with Man*. New York: Plenum.

Seeley, T. 1977. Measurement of nest cavity volume by the honey bee *(Apis mellifera)*. *Behav. Ecol. Sociobiol.* 2:201–227.

Segal, S. J. (Ed.). 1971. *Imagery, Current Cognitive Approaches*. New York: Academic.

Seidenberg, M. S., and Petitto, L. A. 1979. Signing behavior in apes: a critical review. *Cognition* 7:177–215.

Seyfarth, R. M., Cheney, D. L., and Marler, P. 1980a. Vervet monkey alarm calls; semantic communication in a free-ranging primate. *Anim. Behav.* 28:1070–1094.

———. 1980b. Monkey responses to three different alarm calls: evidence of predator classification and semantic communication. *Science* 210:801–803.

Sheehan, P. W. (Ed.). 1972. *The Function and Nature of Imagery*. New York: Academic.

Shepard, R. N. 1978. The mental image. *Am. Psychol.* 33:125–137.

Shorey, H. H. 1977. Pheromones. In: Sebeok, T. A. (Ed.), *How Animals Communicate*. Bloomington: Indiana Univ. Press.

Simmons, J. A., Howell, D. J., and Suga, N. 1975. The information content of bat sonar echoes. *Am. Sci.* 63:204–215.

Simpson, G. G. 1964. *Biology and Man*. New York: Harcourt Brace & World, Chapter 8.

Skinner, B. F. 1957. *Verbal Behavior*. New York: Appleton-Century-Crofts.

———. 1974. *About Behaviorism*. New York: Random House.

Skutch, A. F. 1976. *Parent Birds and their Young*. Austin, Texas: Univ. of Texas Press.

"Smith, Adam." 1972. *Supermoney*. New York: Popular Library.

———. 1975. *Powers of Mind*. New York: Random House.

Smith, K. 1969. *Behavior and Conscious Experience, A Conceptual Analysis*. Athens: Ohio Univ. Press. Smith advocates a form of "physical monism" according to which ". . . awareness itself is a proper part of nature." Conscious experience is defined as "an internal event to which an arbitrary response can be attached directly, by the process of following that response, when it occurs to that event, with a prespecified set of circumstances." Smith believes, despite the behaviorists, that psychology is "the science of the nature, causes, and effects of conscious experience— conscious experience being specified always in terms of the stimulus situations instigating it." He deplores "the somewhat remarkable insistence of modern psychology on avoiding the whole topic of conscious experience . . ." and he advocates instead what he calls "introspectional behaviorism."

Smith, W. J. 1968. Message-Meaning Analysis. In: Sebeok, T. A. (Ed.), *Animal Communication*. Bloomington: Indiana Univ. Press.

———. 1969. Messages of vertebrate communication. *Science* 165:145–150.

———. 1975. Communication, animal. *Encyclopedia Britannica, Macropedia*. Vol. 4, pp. 1010–1019. Chicago: Encyclopedia Britannica.

———. 1977. *The Behavior of Communicating: An Ethological Approach*. Cambridge: Harvard Univ. Press.

Sperry, R. W. 1969. A modified concept of consciousness. *Psychol. Rev.* 76:532–536. (Also discussion with D. Bindra, ibid. 77:581–590, 1970.)

———. 1973. Lateral Specialization of Cerebral Function in the Surgically Separated Hemispheres. In: McGuigan, F. J., and Schoonover, R. A. (Eds.), *The Psychophysiology of Thinking, Studies of Covert Processes*. New York: Academic. This paper and the group discussion published with it demonstrate that both hemispheres of the human cerebral cortex are capable of complex mental processing, and that while normally the dominant hemisphere (usually the left) handles linguistic processes, the division of labor is variable among individuals and probably not sharp and total.

Spuhler, J. N. 1977. Biology, speech and language. *Annu. Rev. Anthropol.* 6:509–561.

Stenhouse, B. 1973. *The Evolution of Intelligence*. London: George Allen and Unwin.

Stevenson, J. C. 1969. Song as a Reinforcer. In: Hinde, R. A. (Ed.), *Bird Vocalizations*. London: Cambridge Univ. Press.

Stout, J. F. 1975. Aggressive communication by *Larus glaucescens*. Part III. Description of the displays related to territorial protection. *Behaviour* 55:181–208.

Stout, J. F., and Brass, M. E. 1969. Aggressive communication by *Larus glaucescens*. Part II. Visual communication. *Behaviour* 34:42–54.

Stout, J. F., Wilcox, C. R., and Creitz, L. E. 1969. Aggressive communication by *Larus glaucescens*. Part I. Sound communication. *Behaviour* 34:29–41.

Straub, R. O., Seidenberg, M. S., Terrace, H. S., and Bever, T. G. 1979. Serial learning in the pigeon. *J. Exp. Anal. Behav.* 32:137–148.

Struhsaker, T. 1967. *Behavior and Ecology of Red Colobus Monkeys*. Chicago: Univ. of Chicago Press.

Suga, N., and O'Neill, W. E. 1979. Neural axis representing target range in the auditory cortex of the mustache bat. *Science* 206:351–353.

———. 1980. Auditory Processing of Echoes: Representation of Acoustic Information from the Environment in the Bat Cerebral Cortex. In: Busnel, R.-G., and Fish, J. F. (Eds.), *Animal Sonar Systems*. New York: Plenum.

Tavolga, W. N. 1974. Application of the Concept of Levels of Organization to the Study of Animal Communication. In: Krames, L., Pliner, P., and Alloway, T. (Eds.), *Nonverbal Communication*. New York: Plenum.

Tax, S., and Callender, C. (Eds.). 1960. *Evolution after Darwin*. Vol. III. *Issues in Evolution*. Chicago: Univ. of Chicago Press.

Taylor, J. G. 1962. *The Behavioral Basis of Perception*. New Haven: Yale Univ. Press. In his final chapter, Taylor concludes that there is no need to postulate any special essence to explain consciousness. "Thus the revolt of the early behaviorists against the 'mentalist' assumptions of classical psychology has led, through the assiduous investigation of the laws of behavior, to a new assertion of the scientific respectability of the once despised phenomena of consciousness."

Teng, E. L., and Sperry, R. W. 1973. Interhemispheric interaction during simultaneous bilateral presentation of letters or digits in commisurotomized patients. *Neuropsychologia* 11:131–140.

Terrace, H. 1979. *Nim*. New York: Knopf.

Terrace, H., Petitto, L. A., Sanders, R. J., and Bever, T. G. 1979. Can an ape create a sentence? *Science* 206:891–902.

Terwilliger, R. F. 1968. *Meaning and Mind, a Study of the Psychology of Language*. New York: Oxford Univ. Press. This book is an example of the viewpoint that ". . . the nature of one's language *determines* his entire way of life, including his thinking and all other forms of mental activity."

Thass-Thienemann, T. 1968. *Symbolic Behavior*. New York: Washington Square Press.

Thatcher, R. W., and John, E. R. 1977. Foundations of Cognitive Processes. Hillsdale, N.J.: Erlbaum.

Thines, G. 1977. *Phenomenology and the Science of Behavior; An Historical and Epistemological Approach*. London: Allen and Unwin.

Thomas, L. 1974. *The Lives of a Cell, Notes of a Biology Watcher*. New York: Viking.

Thorpe, W. H. 1972a. Duetting and antiphonal signing in birds. Its extent and significance. *Behaviour (Suppl.)* 18:1–197.

———. 1972b. In: Hinde, R. A. (Ed.), *Nonverbal Communication*. London: Cambridge Univ. Press, Chapters 2, 5, and 6.

————. 1974a. *Animal Nature and Human Nature*. Garden City, N.Y.: Doubleday.

————. 1974b. Reductionism in Biology. In: Ayala, F. J., and Dobzhansky, T. (Eds.), *Studies in the Philosophy of Biology, Reduction and Related Problems*. Berkeley: Univ. of California Press, Chapter 8.

Tinbergen, N. 1951. *The Study of Instinct*. London: Oxford Univ. Press.

Tolman, E. C. 1932. *Purposive Behavior in Animals and Men*. New York: Century.

————. 1948. Cognitive maps in rats and men. *Psychol. Rev.* 55:189–208.

Turner, L. W. 1973. Vocal and escape responses of *Spermophilus beldingi* to predators. *J. Mammal.* 54:990–993.

Umiker-Sebeok, J., and Sebeok, T. A. 1980. Clever Hans and Smart Simians. The Self-Fulfilling Prophecy and Kindred Methodological Pitfalls. In: Sebeok, T. A., and Umiker-Sebeok, J. (Eds.), *Speaking of Apes, A Critical Anthology of Two-Way Communication with Man*. New York: Plenum.

Underwood, G., and Stevens, R. 1979. *Aspects of Consciousness*. Vol. 1. *Psychological Issues*. New York: Academic.

Walcott, C., and Green, R. P. 1974. Orientation of homing pigeons altered by a change in the direction of an applied magnetic field. *Science* 184:180–182.

Walcott, C., Gould, J. L., and Kirschvink, J. L. 1979. Pigeons have magnets. *Science* 205:1027–1029.

Ward, P., and Zahavi, A. 1973. The importance of certain assemblages of birds as "information centres" for food finding. *Ibis* 115:517–534.

Warren, J. M. 1976. Tool Use in Mammals. In: Masterton, R. B., Campbell, C. B. G., Bitterman, M. E., and Hotton, N. (Eds.), *Evolution of Brain and Behavior in Vertebrates*. Hillsdale, N.J.: Erlbaum.

Waterman, T. H. 1974. The Optics of Polarization Sensitivity. In: Snyder, A. W., and Menzel, R. (Eds.), *Photoreceptor Optics*. New York: Springer.

Watson, J. B. 1929. *Psychology from the Standpoint of a Behaviorist*. Philadelphia: Lippincott.

Weiner, B. 1972. *Theories of Motivation, from Mechanism to Cognition*. Chicago: Markham Publishing Co. In a section on Cognitive Psychology of Infrahumans, Weiner states that "it is reasonable to presume that lower organisms also have cognitions. In infrahumans, however, cognitions are not directly assessible via introspective reports. But they may be inferred from behavior. . . . Morgan's Canon should be considered a guiding statement, rather than an invariant law."

Weiss, D. D. 1975. Professor Malcolm on animal intelligence. *Philos. Rev.* 84:88–95.

Wells, P. H. 1973. Honey Bees. In: Corning, W. C., Dyal, J. A., and Willows, A. O. D. (Eds.), *Invertebrate Learning*. Vol. 2. New York: Plenum, Chapter 9. Wells dismisses completely the evidence that bee dances serve any communicatory function.

Wells, P. H., and Wenner, A. M. 1973. Do honey bees have a language?

Nature (Lond.) 241:171–175.

Wenner, A. M. 1962. Sound production during the waggle dance of the honey bee. *Anim. Behav.* 10:79–95.

———. 1971. *The Bee Language Controversy.* Boulder, Colo.: Educational Programs Improvement Corp.

———. 1974. Information Transfer in Honey Bees: A Population Approach. In: Krames, L., Pliner, P., and Alloway, T. (Eds.), *Nonverbal Communication.* New York: Plenum.

Westby, G. W. M. 1974. Assessment of the signal value of certain discharge patterns in the electric fish, *Gymnotus carapo*, by means of playback. *J. Comp. Physiol.* 92:327–341.

———. 1975a. Comparative studies of the aggressive behaviour of two gymnotid electric fish *(Gymnotus carapo* and *Hypopomus artedi). Anim. Behav.* 23:192–213.

———. 1975b. Further analysis of the individual discharge characteristics predicting social dominance in the electric fish *Gymnotus carapo. Anim. Behav.* 23:249–260.

———. 1975c. Has the latency dependent response of *Gymnotus carapo* to discharge-triggered stimuli a bearing on electric fish communication? *J. Comp. Physiol.* 96:307–341.

Whitehead, A. N. 1938. *Modes of Thought.* New York: Macmillan.

Wilcox, R. S. 1972. Communication by surface waves: Mating behavior of a water strider (Gerridae). *J. Comp. Physiol.* 80:225–266.

Wilkinson, D. H. 1952. The random element in bird "navigation." *J. Exp. Biol.* 29:532–560.

Wilson, E. O. 1971. *The Insect Societies.* Cambridge: Harvard Univ. Press.

———. 1975. *Sociobiology, the New Synthesis.* Cambridge: Harvard Univ. Press.

Wiltschko, W. 1974. Der Magnetkompass der Gartengrassmucke *(Sylvia borin). J. Ornithol.* 115:1–7.

———. 1975. The interaction of stars and magnetic field in the orientation system of night migrating birds. *Z. Tierpsychol.* 37:337–355.

Wittgenstein, L. 1953. *Philosophical Investigations.* 3rd ed. (Translated by Anscombe, G.E.M.). New York: Macmillan.

Wood, S., Moriarty, K. M., Gardner, B. T., and Gardner, R. A. 1980. Object Permanence in Child and Chimpanzee. *Anim. Learn. Behav.* 8:3–9.

Yerkes, R., and Yerkes, A. W. 1929. *The Great Apes, a Study of Anthropoid Life.* New Haven: Yale Univ. Press.

Subject Index

Author Index